Contents

Xinjiang Uygur Zizhiqu
Gansu
Qinghai
Ningxia
Huizu
Zizhiqu
Shaanxi
Xizang Zizhiqu
Huang He
Sichuan
Chongqing
Jinsha Jiang
Guizhou
Yunnan

Beijing
Urban Area of Beijing
Tianjin
Hebei Province
Shanxi Province
Nei Mongol Zizhiqu

Henan Province
Hubei Province
Hunan Province
Guangdong Province
Guangxi Zhuangzu Zizhiqu
Hainan Province

Liaoning Province
Jilin Province
Heilongjiang Province

Chongqing
Sichuan Province
Guizhou Province
Yunnan Province
Xizang Zizhiqu

Shanghai
Jiangsu Province
Zhejiang Province
Anhui Province
Fujian Province
Jiangxi Province
Shandong Province

Shaanxi Province
Gansu Province
Qinghai Province
Ningxia Huizu Zizhiqu
Xinjiang Uygur Zizhiqu

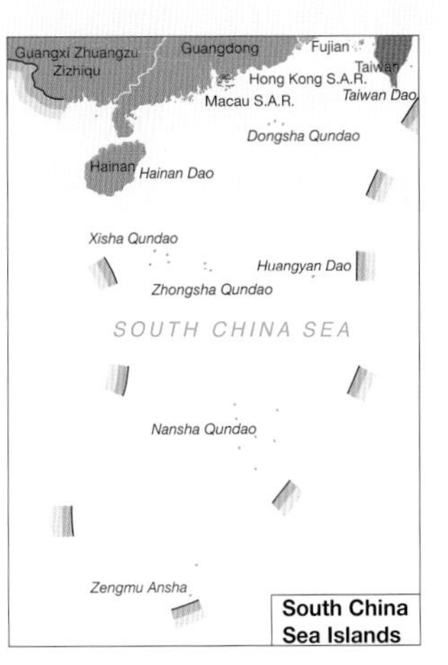

Hong Kong S.A.R.
Macau S.A.R.
Taiwan Province
Diaoyu Dao
South China Sea Islands

How to Use the Book

(Basic Structure of the Book with Hainan Province as an example)

Location on Map

To indicate the location of the region in China

Illustration and Explanation

To demonstrate major attractions, significant characters, unique architecture, culture and customs, local products, rare species, pillar industries and so on

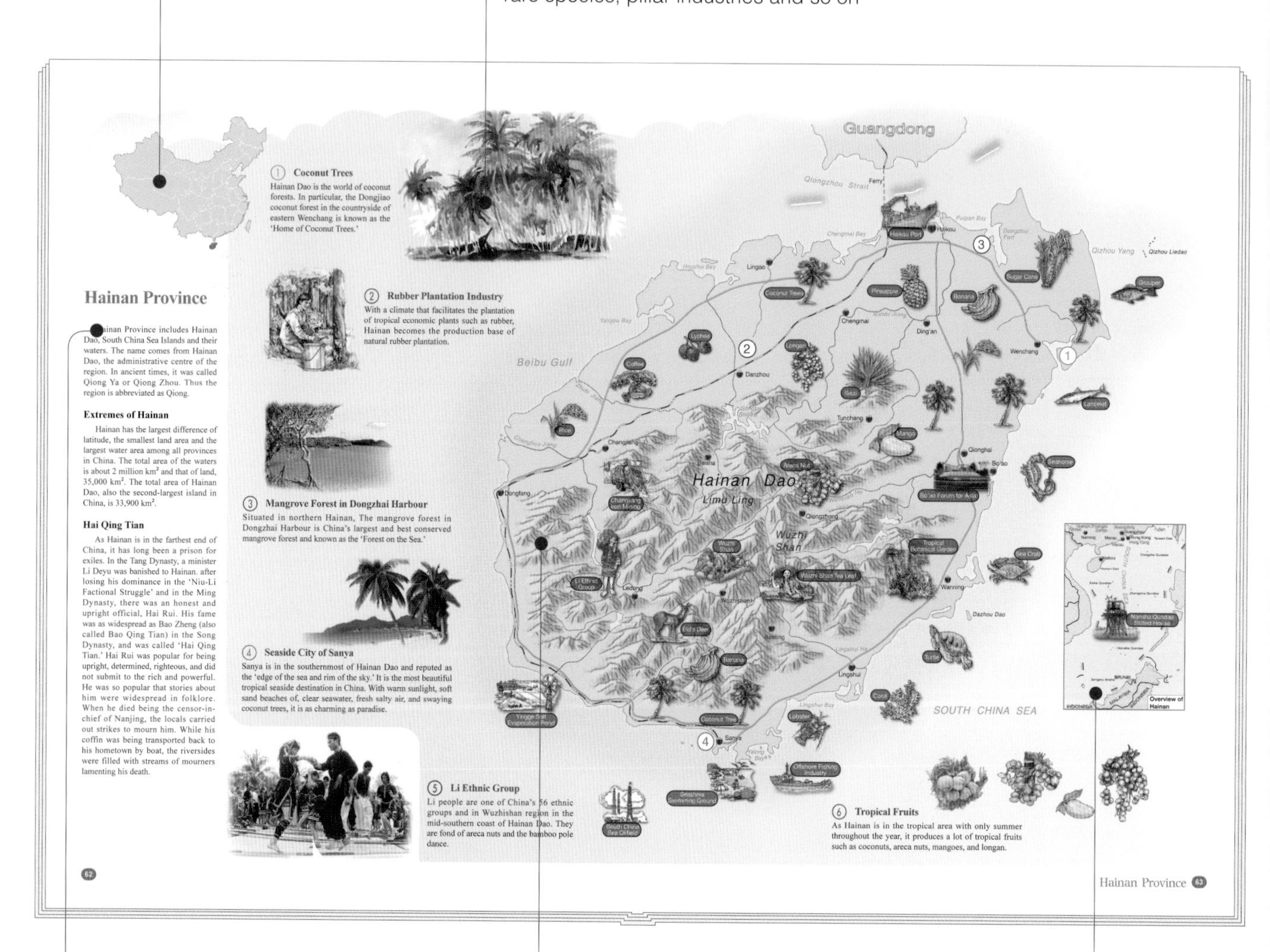

Hainan Province

Hainan Province includes Hainan Dao, South China Sea Islands and their waters. The name comes from Hainan Dao, the administrative centre of the region. In ancient times, it was called Qiong Ya or Qiong Zhou. Thus the region is abbreviated as Qiong.

Extremes of Hainan

Hainan has the largest difference of latitude, the smallest land area and the largest water area among all provinces in China. The total area of the waters is about 2 million km^2 and that of land, 35,000 km^2. The total area of Hainan Dao, also the second-largest island in China, is 33,900 km^2.

Hai Qing Tian

As Hainan is in the farthest end of China, it has long been a prison for exiles. In the Tang Dynasty, a minister Li Deyu was banished to Hainan. after losing his dominance in the 'Niu-Li Factional Struggle' and in the Ming Dynasty, there was an honest and upright official, Hai Rui. His fame was as widespread as Bao Zheng (also called Bao Qing Tian) in the Song Dynasty, and was called 'Hai Qing Tian.' Hai Rui was popular for being upright, determined, righteous, and did not submit to the rich and powerful. He was so popular that stories about him were widespread in folklore. When he died being the censor-in-chief of Nanjing, the locals carried out strikes to mourn him. While his coffin was being transported back to his hometown by boat, the riversides were filled with streams of mourners lamenting his death.

① **Coconut Trees**

Hainan Dao is the world of coconut forests. In particular, the Dongjiao coconut forest in the countryside of eastern Wenchang is known as the 'Home of Coconut Trees.'

② **Rubber Plantation Industry**

With a climate that facilitates the plantation of tropical economic plants such as rubber, Hainan becomes the production base of natural rubber plantation.

③ **Mangrove Forest in Dongzhai Harbour**

Situated in northern Hainan, The mangrove forest in Dongzhai Harbour is China's largest and best conserved mangrove forest and known as the 'Forest on the Sea.'

④ **Seaside City of Sanya**

Sanya is in the southernmost of Hainan Dao and reputed as the 'edge of the sea and rim of the sky.' It is the most beautiful tropical seaside destination in China. With warm sunlight, soft sand beaches of, clear seawater, fresh salty air, and swaying coconut trees, it is as charming as paradise.

⑤ **Li Ethnic Group**

Li people are one of China's 56 ethnic groups and in Wuzhishan region in the mid-southern coast of Hainan Dao. They are fond of areca nuts and the bamboo pole dance.

⑥ **Tropical Fruits**

As Hainan is in the tropical area with only summer throughout the year, it produces a lot of tropical fruits such as coconuts, areca nuts, mangoes, and longan.

62

Hainan Province 63

History and Literature

To introduce the origin of the name of the region, allusions related to the region such as idioms, literature, and historical events

Regional Map

To display the relief of natural landscape of the region and to show the distribution of major cities, transportation networks, tourist destinations, and local specialities

Supplementary Map

To present distric related to but not within the regiona map

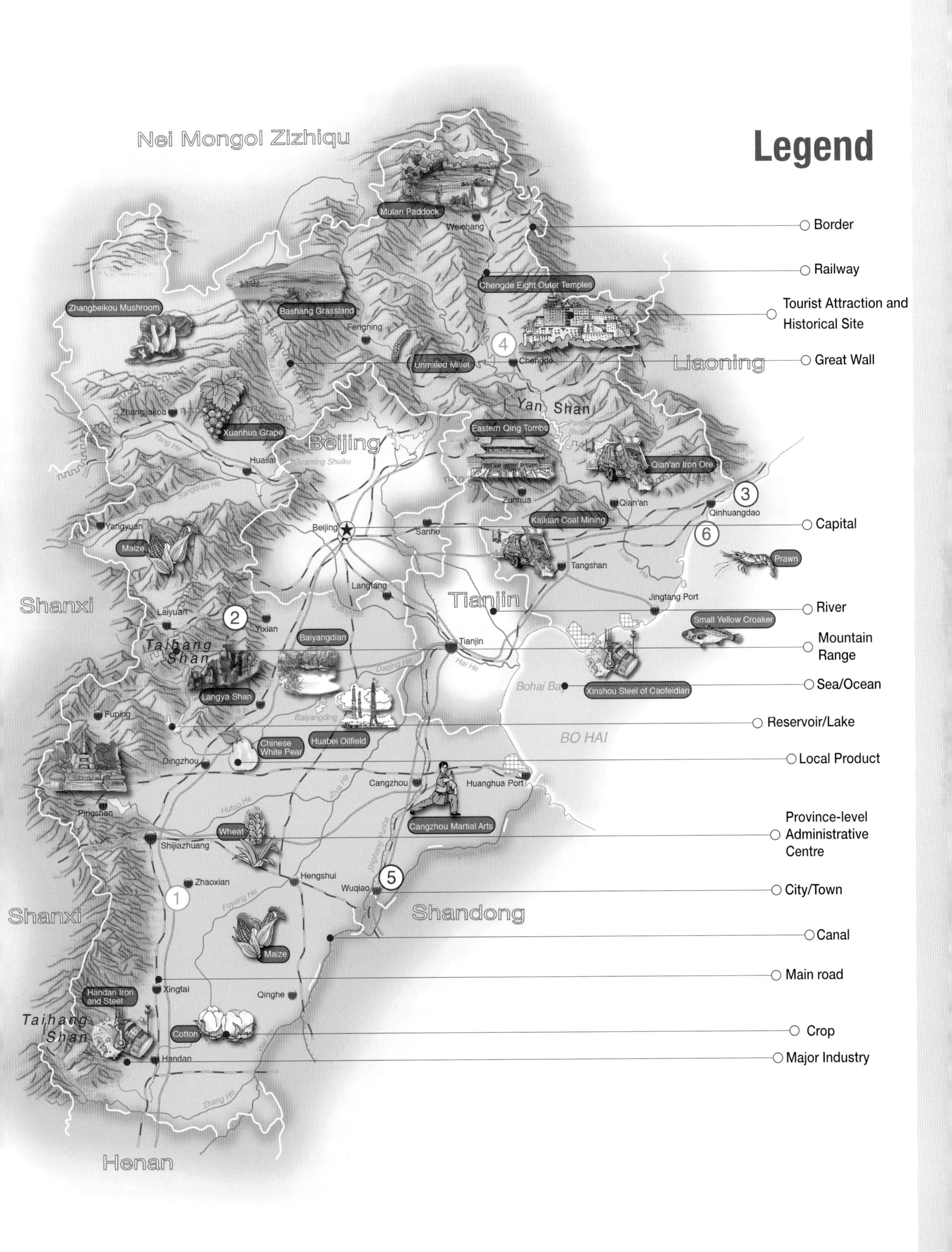
Legend
Border
Railway
Tourist Attraction and Historical Site
Great Wall
Capital
River
Mountain Range
Sea/Ocean
Reservoir/Lake
Local Product
Province-level Administrative Centre
City/Town
Canal
Main road
Crop
Major Industry
Nei Mongol Zizhiqu
Liaoning
Shanxi
Shandong
Henan
Beijing
Tianjin
Mulan Paddock
Weichang
Chengde Eight Outer Temples
Zhangbeikou Mushroom
Bashang Grassland
Fengning
Unmilled Millet
Chengde
Yan Shan
Zhangjiakou
Xuanhua Grape
Eastern Qing Tombs
Qian'an Iron Ore
Huailai
Guanting Shuiku
Zunhua
Qian'an
Qinhuangdao
Kailuan Coal Mining
Yangyuan
Maize
Sanhe
Tangshan
Prawn
Langfang
Laiyuan
Yixian
Jingtang Port
Small Yellow Croaker
Taihang Shan
Baiyangdian
Langya Shan
Daqing He
Hai He
Bohai Bay
Xinshou Steel of Caofeidian
Fuping
Baiyangding
Chinese White Pear
Huabei Oilfield
BO HAI
Dingzhou
Cangzhou
Huanghua Port
Pingshan
Hutuo He
Ziya He
Cangzhou Martial Arts
Wheat
Shijiazhuang
Hengshui
Zhaoxian
Wuqiao
Fuyang He
Xingtai
Qinghe
Handan Iron and Steel
Cotton
Handan
Zhang He
1
2
3
4
5
6

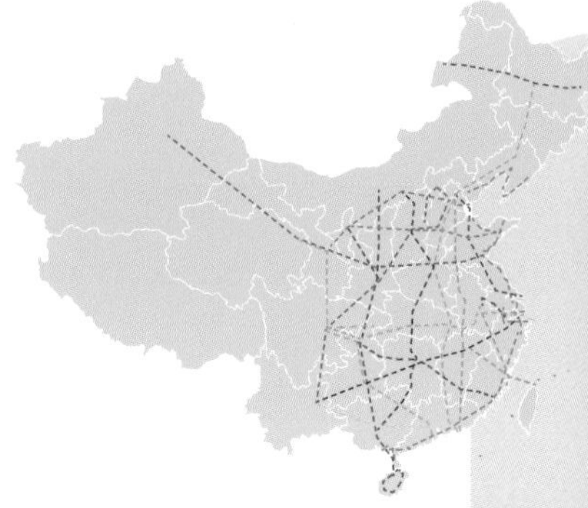

China's Development and Take-off

In the first 20 years of the 21st century, China has become the world's second largest economic entity and advocated many initiatives and concepts of development, including 'The Belt and Road', 'The Great Rejuvenation of the Chinese Nation', 'China Dream', 'Building a Community with a Shared Future for Mankind'.

From 2012 to 2017, China has made remarkable achievements in many fields, such as economy, technology, and eco-development, etc.

Tianyan in China

The Five-Hundred-Metre Aperture Spherical Radio Telescope (FAST), nicknamed Tianyan (Eye of the Sky), is located in Pingtang County, Guizhou, southwest China. It was put into service on 25th September 2016. FAST is the world's largest and most sensitive filled-aperture radio telescope, over which China owns independent intellectual property rights.

Hydraulic Engineering

Xinjiang Uygur Zizhiqu

China's installed hydroelectric capacity reached 330 GW by the end of 2016, remaining the first in the world. The Xiluodu Dam on the Jinsha River, the upper course of the Yangtze, was put into service in 2014, ranking the second largest hydroelectric power station in China, and the third in the world.

Aircraft Carrier Liaoning

The first aircraft carrier commissioned into the People's Liberation Army Navy (PLAN), was originally constructed for the Soviet Navy as *Varyag*, purchased and rebuilt by China in 1999, and commissioned into the PLAN on 25th September 2012.

Comac C919

C919 is a jet airliner independently developed and manufactured by China, able to carry 158—168 passengers from 4,075 up to 5,555 km. The airliner's first flight was undertaken on 5th May 2017.

Satellite Mozi (*Micius*)

China's first Quantum Experiments at Space Scale (QUESS), nicknamed Mozi (*Micius*) after the ancient Chinese philosopher and scientist, was launched on 15th September 2016. Mozi (*Micius*) made China the first country in realising the quantum network between the Earth and satellites, ensuring that communication is safeguarded against being tapped.

Ecology and Greening of Lands

Within 5 years, China has afforested more than 450 million *mu* (1 *mu* equals 667 m^2) of lands and conserved 600 million *mu* of forests, witnessing the most rapid increase in forest resources globally.

High Speed Railway Network

China has built the largest and fastest high speed railway (HSR) network with complete independent intellectual property rights. Up till now, the network, has extended the "Four Vertical and Four Horizontal" network to a new "Eight Vertical and Eight Horizontal" network, consisting of eight north-south (vertical) corridors and eight east-west (horizontal) ones. The HSR mainlines in the "Eight + Eight" passageway grid are built to accommodate train speeds of 250 to 350 km/h.

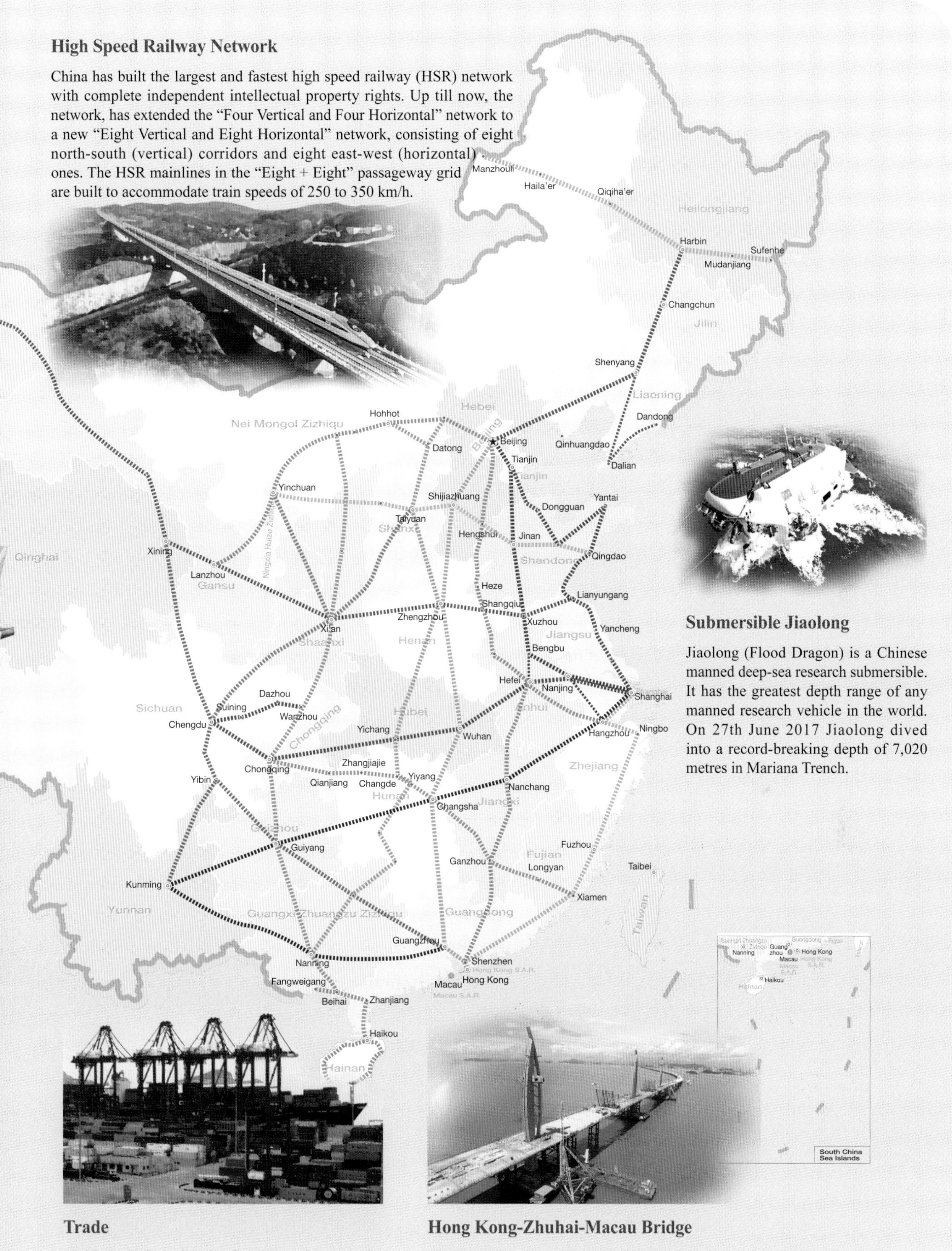

Submersible Jiaolong

Jiaolong (Flood Dragon) is a Chinese manned deep-sea research submersible. It has the greatest depth range of any manned research vehicle in the world. On 27th June 2017 Jiaolong dived into a record-breaking depth of 7,020 metres in Mariana Trench.

Trade

In 2013, China, for the first time, became the largest trading nation in the world. As the leading exporting country which ranks the second in imports, China's export market share maintains higher than 13%.

Hong Kong-Zhuhai-Macau Bridge

The longest sea crossing bridge on earth has a total length of 55 kilometres. It connects Hong Kong, Zhuhai, and Macau. The whole structure of the bridge was completed in July 2017. Its immersed tunnel, deeply sunk into the seabed, is the longest and the only one of its kind in the world.

Territory of China

Historical China

Ancient China was one of the oldest civilizations and birthplaces of modern civilizations, together with ancient Egypt, Babylon, India, and Greece. Just like other ancient civilizations, China started from a primitive tribe covering a small area and finally developed into a large country through continuous unification and expansion.

Across Chinese history, such dynasties as Han, Tang, Yuan, Ming, and Qing built powerful empires with expansive territories and a huge population. But the territories of these empires could be quite different from the territory of modern China, which is roughly based on that of Ming and Qing dynasties.

As compared to ancient Egypt, Babylon, and India, China is a continuous civilization. Modern people find it difficult to understand their ancestry as they cannot understand ancient words. However, modern Chinese can read oracles which are Chinese characters from thousands of years ago and thus know about Chinese civilization.

Location of China on the Earth

From a hemispherical view of the Earth, China is in the Northern Hemisphere and Eastern Hemisphere. For continental and coastal location, it is in the eastern part of Eurasia and the west of the Pacific Ocean. It is bordered by both land and seas.

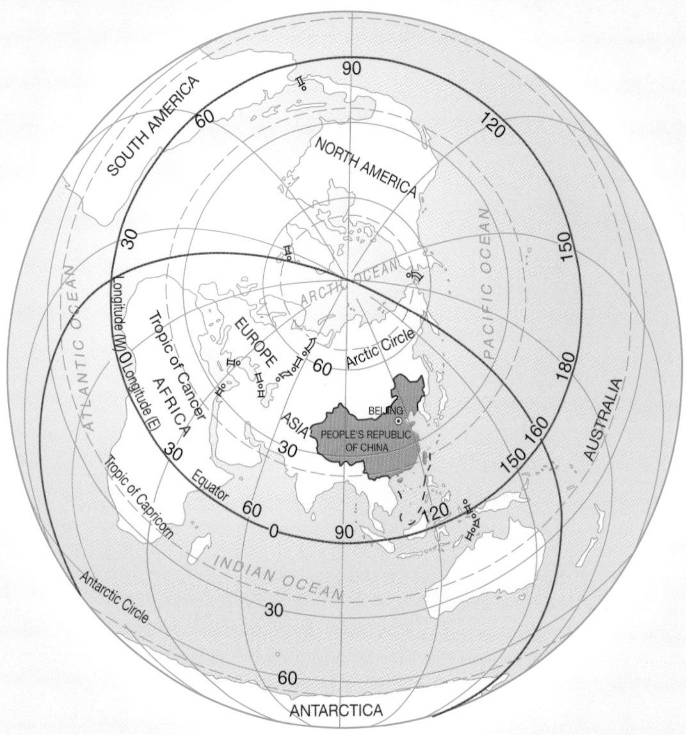

Location of China on the Earth

Territory of China

China is a large country with vast expanses of land and sea. Spanning 9.6 million km^2 on land, it is the world's third largest country just behind Russia and Canada. With about 3 million km^2 of waters, China is also one of the few countries with such a large domain on seas. China has a long land perimeter of over 20,000 km and a coastal perimeter of over 18,000 km.

Four Boundary Points of Border of China

The furthermost points in China in the four directions, often called the four boundary points, are the eastern point at the intersection of Heilong Jiang and the median line of the main channel of Wusuli Jiang, the western point at Pamir Gaoyuan, the northern point at the median line of the main channel of Heilong Jiang, and the southern point at Zengmu Ansha in Nansha Qundao.

Pamir Gaoyuan

Wusuli Jiang

Dawn and dusk in the easternmost and westernmost points in China

There are a 62-degree difference in longitudes and four-hour time lag between the easternmost and westernmost points of China—when the morning glow glitters on the Wusuli Jiang, the starry night sky is still hanging over the Pamir Gaoyuan.

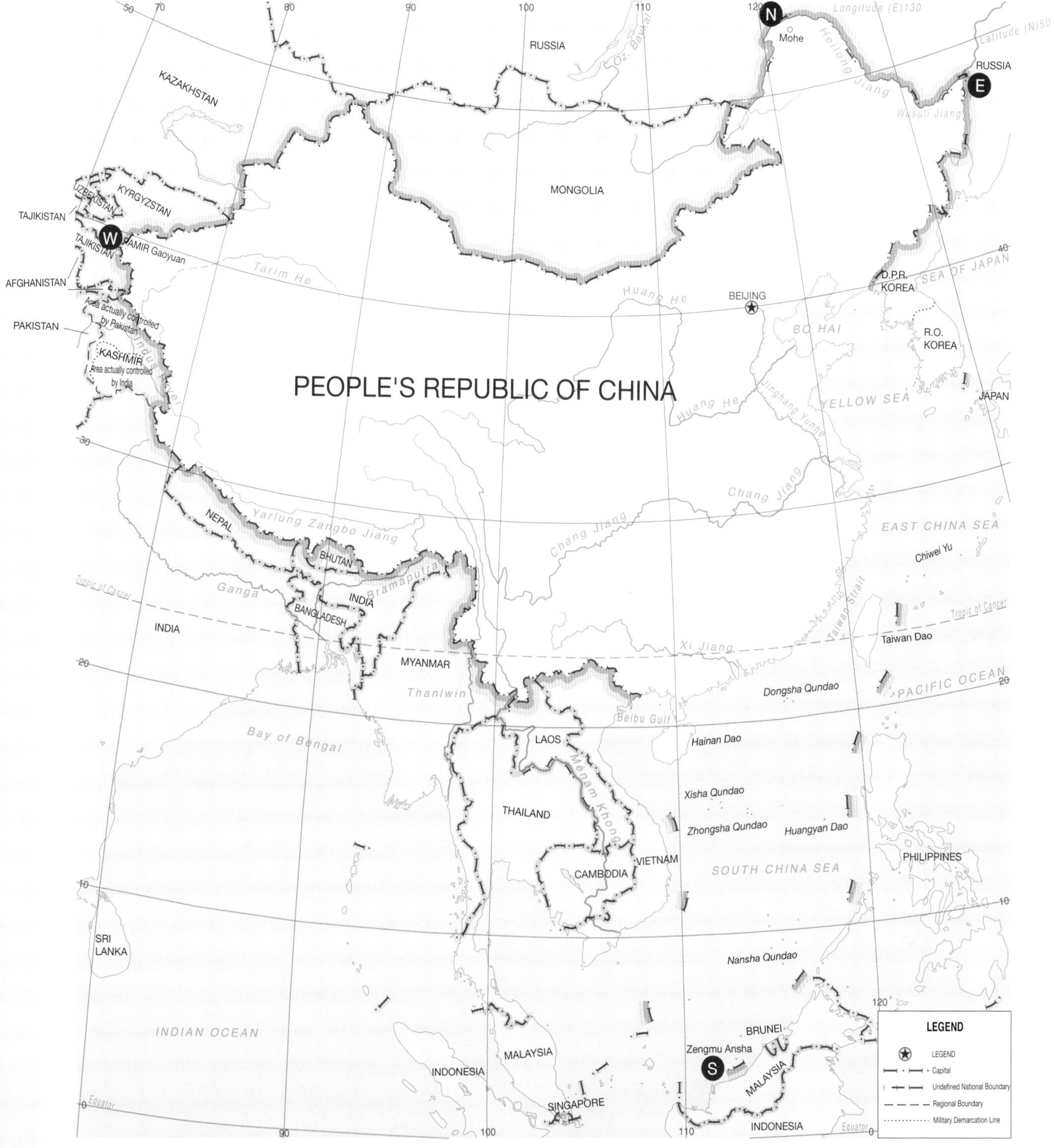

Territory of China

Hainan Dao

Heilongjiang

Seasonal Variation in Southern and Northern China

There is a roughly 50-degree difference in latitudes of northern and southern China. When Hainan Dao is ready for spring planting, Heilongjiang is still wrapped in snow.

China's Neighbours on Land

China is adjacent to 20 countries, 14 of which are bordered on land: D. P. R. Korea, Russia, Mongolia, Kazakhstan, Kyrgyzstan, Tajikistan, Afghanistan, Pakistan, India, Nepal, Bhutan, Myanmar, Laos, and Vietnam.

China's Neighbouring Countries

China has been closely related to its neighbouring countries across history. In the Han Dynasty, diplomats such as Zhang Qian and Ban Chao were assigned to visit the Western Regions, reaching the present-day Afghanistan and Uzbekistan in Central Asia. In the Tang Dynasty, the eminent monk Xuan Zhuang went east to India to study Buddhism and Jian Zhen went west to Japan to spread Buddhism; Japan also sent special envoys to study in China. In 1417 A.D. (Ming), the Eastern King of Sulu (present-day Sulu Archipalego in the Philippines) brought a large delegation to China. They were received cordially by the Emperor Chengzu of Ming. Unfortunately, the Eastern King passed away in China due to illness. He was buried in Dezhou, Shandong. China is also closely related to other countries such as Vietnam and those in the Korean peninsula.

China's Neighbours on Land

The Japanese Official in China

Chao Heng, originally named Abe no Nakamaro, was appointed in 716 A.D. (Tang) by the Japanese government to study in China. He then became an official in the Tang Dynasty and a close friend with famous poets such as Li Bai. In 753 A.D. (Tang) when Chao was sailing back to Japan, he was rumoured dead on the voyage. Li was grieved hearing the news, then wrote a mourning poem Ku Chao Qing Heng in deep sorrow.

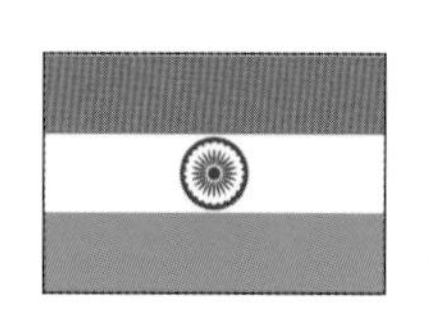
India

China's Neighbours at Sea

China is bordered by 8 countries at sea: D. P. R. Korea, R. O. Korea, Japan, the Philippines, Malaysia, Brunei, Indonesia, and Vietnam. Among them, North Korea and Vietnam are connected to China both on land and at sea, while the other 6 countries are separated from China by seas.

China's Neighbours at Sea

D. P. R. Korea

R.O. Korea

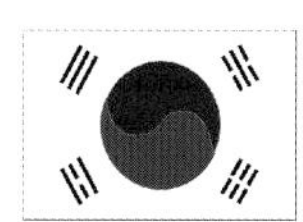
Japan

The Philippines

Malaysia

Brunei

Indonesia

Vietnam

The Gate of Manchuria at the Sino-Russian Border

Chinese Soldiers Guarding the Nansha Qundao

Provincial Divisions in China

Administrative regions in China are divided into three levels: provinces (autonomous regions, municipalities, and special administrative regions), counties (autonomous counties and cities), and townships (autonomous townships and towns). Currently China has 34 provincial administrative units comprising 23 provinces, 5 autonomous regions, 4 municipalities, and 2 special administrative regions.

Administrative Regions in China

The Largest

Xinjiang is the largest province in China with the longest land perimeter and bordered with the most countries. It has a total area of 1.6649 million km² which equals 1/6 of the land in China. Bordered with 8 countries, it has a land perimeter of more than 5,600 km which equals 1/4 of the total land perimeter of China.

The Smallest

Macau is of 32.9 km², roughly 1/34 of Hong Kong and 1/22 of Singapore.

The Most Populated

Up till 29 April 2011, the five most populated regions are: Guangdong, Shandong, Henan, Sichuan, and Jiangsu. Guangdong is the most populated province with 104 million residents, which accounts for 8% of population in China.

The Least Populated

Up till 2011, Macau Special Administrative Region has 558,000 residents. It is the least populated province-level region in China. Xizang is the least populated province with 3.0022 million residents with the lowest population density of less than 3 people for each km².

National Emblem of the PRC

National Flag of the PRC

List of Province-level Administrative Units in China

Names	Abbreviations	Administrative Centres
Beijing	Jing (京)	Beijing
Tianjin	Jin (津)	Tianjin
Hebei Province	Ji (冀)	Shijiazhuang
Shanxi Province	Jin (晉)	Taiyuan
Nei Mongol Zizhiqu	Neimenggu (內蒙古)	Hohhot
Liaoning Province	Liao (遼)	Shenyang
Jilin Province	Ji (吉)	Changchun
Heilongjiang Province	Hei (黑)	Harbin
Shanghai	Hu (滬)	Shanghai

Names	Abbreviations	Administrative Centres
Jiangsu Province	Su (蘇)	Nanjing
Zhejiang Province	Zhe (浙)	Hangzhou
Anhui Province	Wan (皖)	Hefei
Fujian Province	Min (閩)	Fuzhou
Jiangxi Province	Gan (贛)	Nanchang
Shandong Province	Lu (魯)	Jinan
Henan Province	Yu (豫)	Zhengzhou
Hubei Province	E (鄂)	Wuhan
Hunan Province	Xiang (湘)	Changsha

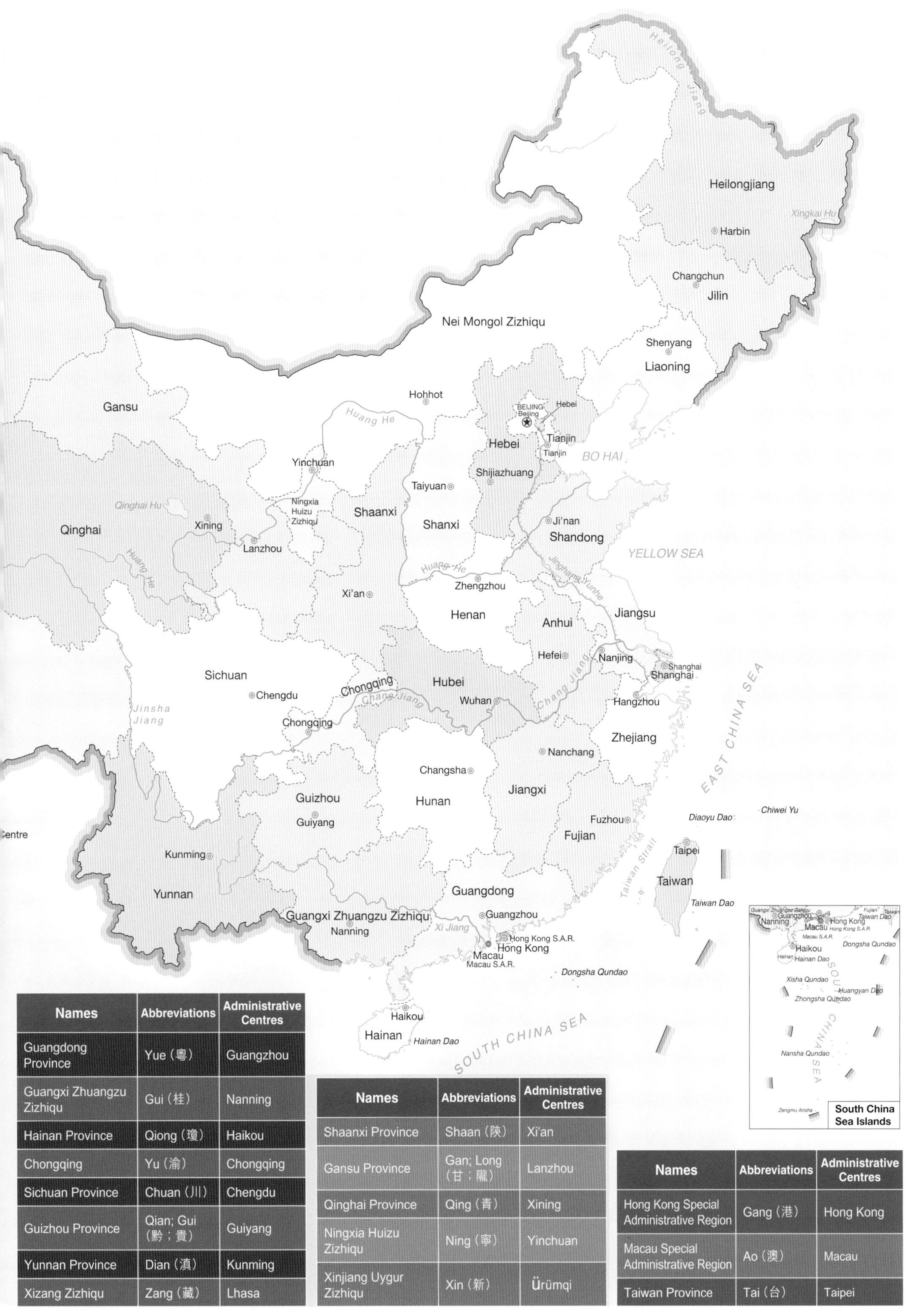

Names	Abbreviations	Administrative Centres
Guangdong Province	Yue (粵)	Guangzhou
Guangxi Zhuangzu Zizhiqu	Gui (桂)	Nanning
Hainan Province	Qiong (瓊)	Haikou
Chongqing	Yu (渝)	Chongqing
Sichuan Province	Chuan (川)	Chengdu
Guizhou Province	Qian; Gui (黔；貴)	Guiyang
Yunnan Province	Dian (滇)	Kunming
Xizang Zizhiqu	Zang (藏)	Lhasa

Names	Abbreviations	Administrative Centres
Shaanxi Province	Shaan (陝)	Xi'an
Gansu Province	Gan; Long (甘；隴)	Lanzhou
Qinghai Province	Qing (青)	Xining
Ningxia Huizu Zizhiqu	Ning (寧)	Yinchuan
Xinjiang Uygur Zizhiqu	Xin (新)	Ürümqi

Names	Abbreviations	Administrative Centres
Hong Kong Special Administrative Region	Gang (港)	Hong Kong
Macau Special Administrative Region	Ao (澳)	Macau
Taiwan Province	Tai (台)	Taipei

A Multi-ethnic Country

There are 56 ethnic groups living on the expansive land area of China, such as Han, Zhuang, Manchu, Hui, Miao, Uygur, Tujia, Yi, Mongol, and Tibetan. Han is the largest group making up 91.6% of the total population. Other 55 groups, usually called the 'minority groups,' account for only 8.4%. They mainly live in the southwest, northwest, and northeast of China. To sum up, Chinese ethnic groups are widely dispersed with certain degree of concentration within each group.

Distribution of Han and Minori Ethnic Groups

Han Ethnic Group 91.6%

Minority Ethnic Groups 8.4%

Distribution of Ethnic Groups in China

The Smallest Ethnic Group

Lhoba is the smallest ethnic group in China with only 3,000 people. They spread in Tibet from Zayü in the east and the Luoyu region in Menyu in the west of Tibet, in particular they live in Milin, Mêdog, Zayü, Longzi, and Langxian. 'Lhoba' is a Tibetan word meaning people of the south. Most of them are farmers and hunters. They have their own spoken language but do not have written language. They are still using the primitive ways of carving on wood and tying ropes to calculate and record.

The Largest Ethnic Group

Zhuang is the largest ethnic group in China with over 17 million people. They mainly live in Guangxi, Yunnan, Guangdong, and Guizhou.

Costumes of Various Ethnic Groups

Han Zhuang Mongol Uygur

Tibetan Hui Chaoxian Kaza

Miao Yi Bouyei Oroqen

Manchu Dong Hani Tajik

Yao Bai Lisu Daur

Li Dai Tujia Ewenki

Dongxiang Mulam Lhoba Tu

Distribution of Ethnic Groups in China

Geographical Landscape of China

General Geographical Features of China

From an aerial view, China is like a stairway descending from west to east. The first step is Qingzang Gaoyuan in the west with an average elevation of over 4,000 m above sea-level. The second step consists of Nei Mongol Gaoyuan, Huangtu Gaoyuan, Yungui Gaoyuan, Tarim Pendi, Junggar Pendi, and Sichuan Pendi with an average elevation of 1,000—2,000 m. The third step starts from the eastern edge of Da Hing'an Ling, to Taihang Shan, Wu Shan, Xuefeng Shan, and finally the coast of the Pacific Ocean and the elevation drops to 500—1,000 m. The fourth step spans from north to south and consists of the Dongbei Pingyuan, Huabei Pingyuan, Changjiang Zhongxiayou Pingyuan, and the neritic zone of China. The neritic zone is less than 200 m below water and a rich source of marine resources.

North-South Boundary of China

The north-south boundary is marked by Qin Ling and Huai He. There is a stark contrast between the two sides in natural resources, agricultural production methods, geographical landscapes, and customs. The north of Qin Ling is often called northern China while the south is called southern China.

Various Natural Landscapes

China has an expansive land area in forms of mountains, plateaus, basins, plains, and hilly land. It demonstrates various kinds of natural landscapes.

Hilly Land 10%

Plain 12%

Basin 19%

Plateau 26%

Mountain 33%

World's Geographical Extremes in China

Highest Plateau:	Qingzang Gaoyuan, 4,500 m above sea-level and nicknamed the 'Roof of the World'
Lowest Basin:	Ayding Hu in Tarim Pendi, 154 m below sea-level and China's lowest point on land
Highest Mountains:	Himalayas, of which 10 peaks are among the world's most famous 14 peaks over 8,000 m above sea-level
Highest Peak:	Qomolangma Feng, 8,844.43 m above sea-level on the Sino-Nepalese boundary
Deepest Valley:	Yarlung Zangbo Grand Canyon, with the greatest depth of 6,009 m and an average depth of 2,268 m
Longest Canal:	Jinghang Yunhe, 1,800 km long from Beijing in the north to Hangzhou in the south

Modern Transport in China

Ancient Transport in China—'Ting' and 'Yizhan'

In ancient times, due to the lack of a good transport network, it was inconvenient to have meals and accommodation when travelling. 'Ting' mentioned in Li Bai's poem *Pusaman* and Master Hong Yi's poem *Songbie*, were houses for people to take rest and built by the government beside the road with certain distance between each one. These houses were first called 'ting.' It is said that a 'changting' was built every 10 *li* and a 'duanting' every 5 *li*. Later, 'changting' became the place of gathering and parting. Thanks to the depiction by generations of literati, 'changting' even became the synonym of a place to bid farewell.

'Yizhan' were postal relay stations for officials delivering documents and messages or for travelling officials to have meals and accommodation and to change horses. The Qin Dynasty stipulated that a 'ting' should be built along the road every 10 *li*, and that the 'ting' should provide accommodation and meals for delivery men on horses. Liu Bang, Emperor Gaozu of Han was a 'ting zhang' (an official working in a 'ting') in the Qin Dynasty.

Modern Transport in China

Transport in China is very well-developed. There are trains and cars on land. Ships sail on seas and rivers, and planes fly in the air. It is developed into an interconnected, comprehensive transport network.

Main Roads and Ports in China

Plane

Ship

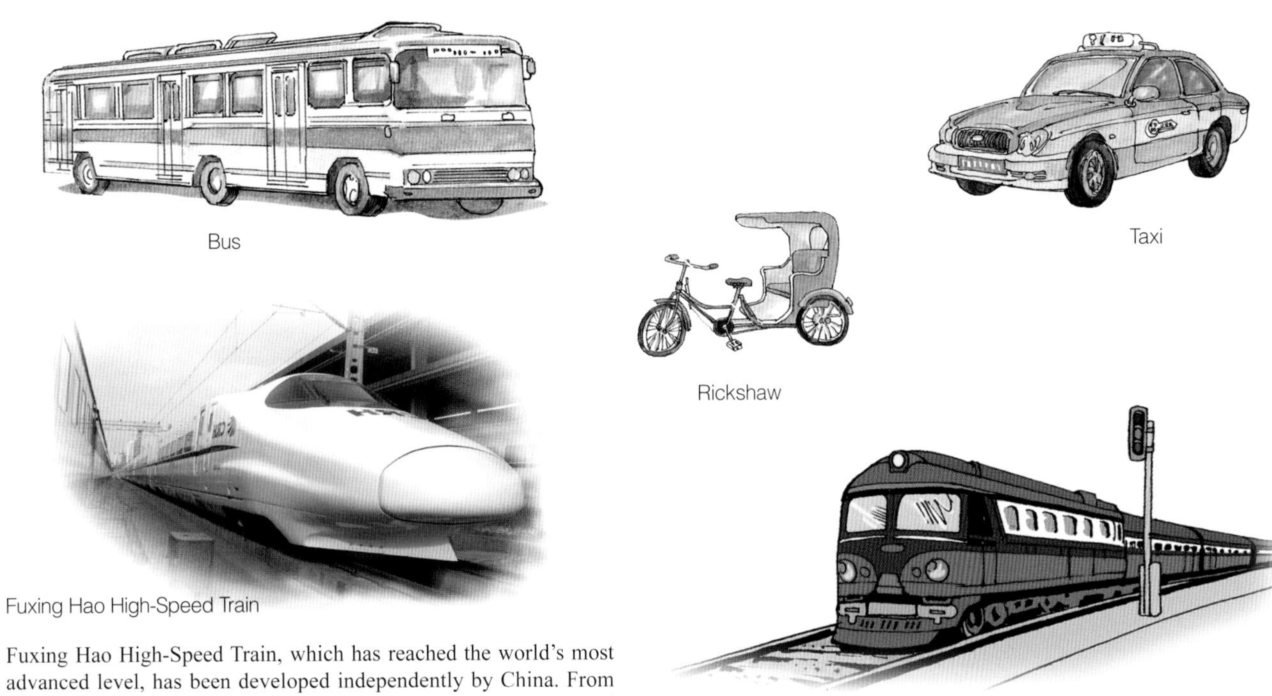

Bus

Taxi

Rickshaw

Fuxing Hao High-Speed Train

Train

Fuxing Hao High-Speed Train, which has reached the world's most advanced level, has been developed independently by China. From 21th August 2017 onwards, Fuxing Hao has been running at a speed of 350 kph on the Beijing-Shanghai Highspeed Railway. China now owns and operates the world's fastest commercial high-speed trains.

Main Railways and Airports in China

Tourism in China

UNESCO World Heritage in China

Among China's 43 UNESCO World Heritage sites approved by the UNESCO committee, 30 out of the 43 sites are cultural heritage sites (3 are cultural landscape and scenery sites), 9 are natural heritage sites, 4 are those sites which contain dual features of cultural and natural heritage. For the time being, China ranks the second in the world in terms of the number of UNESCO World Heritage sites whereas Italy ranks the first.

Abundant Tourist Resources

China is an old and enormous country famous for tourism with widely different natural landscapes, elegant scenery of mountains and water, splendid historical sites, and charming customs of ethnic groups. It has nearly 200 national-level major scenic sites.

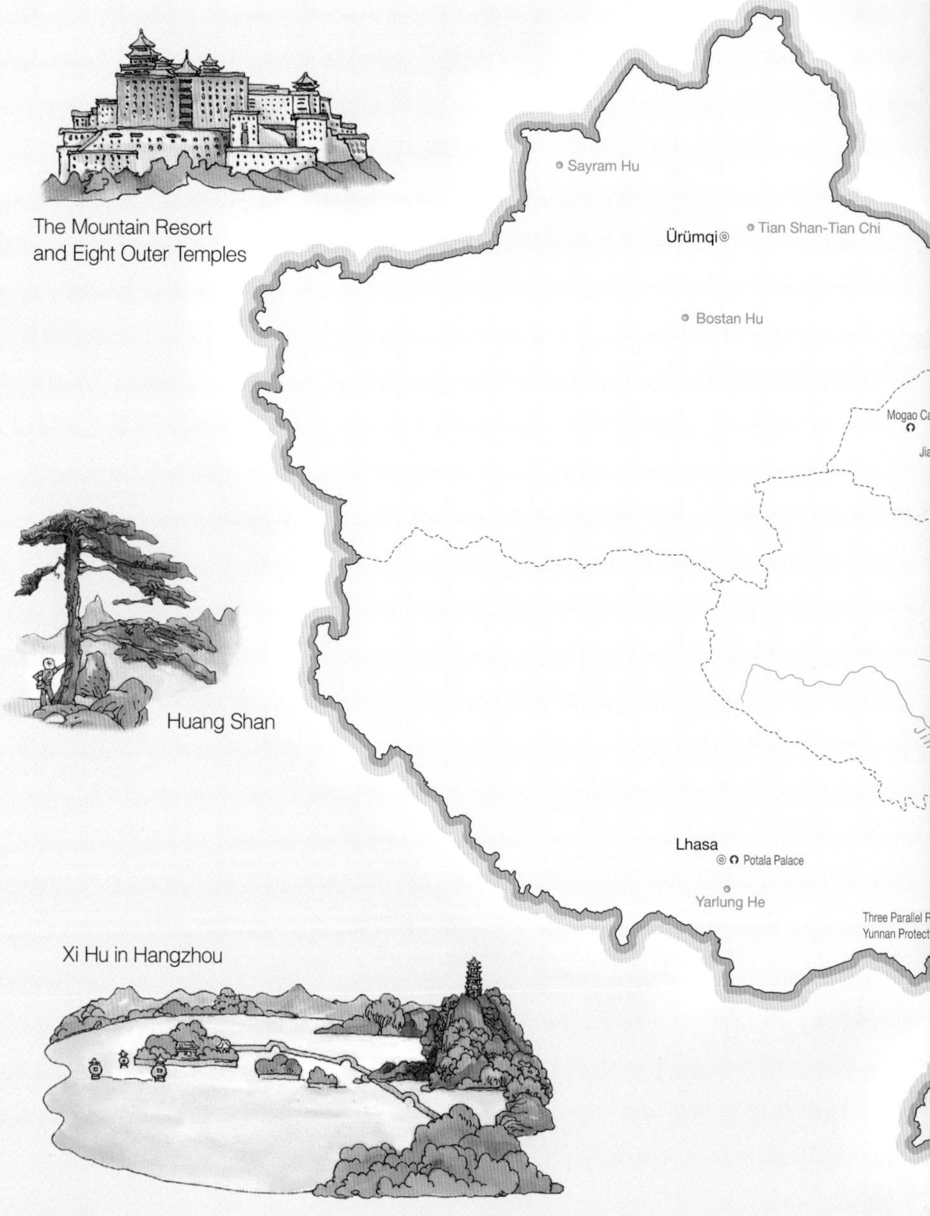

The Mountain Resort and Eight Outer Temples

Huang Shan

Xi Hu in Hangzhou

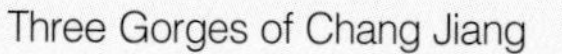

Three Gorges of Chang Jiang

Sun Moon Lake

Landscape of Guilin
Classical Gardens of Suzhou
Imperial Palace of Beijing
Great Wall
Terracotta Army
Heilong Jiang
Wudalianchi
Zalantun
Xingkai Hu
Harbin
Jingpo Hu
Changchun
Songhua Hu
Badabu-Jingyuetan
The Capital Cities and Tombs of the Ancient Koguryo Kingdom
Shenyang Imperial Palace, Wuling Tomb, Fuling Tomb
Shenyang
Yongling Tomb
Qian Shan
Yalu Jiang
Xingcheng Beach
The Mountain Resort and Eight Outer Temples
Hohhot
Badaling Great Wall
Ming Tombs
Eastern Qing Tombs
Huang He
Yungang Grottoes
Beijing
Temple of Heaven, the Palace Museum, Summer Palace
The Mountain Resort and Eight Outer Temples
Qinhuang-Dao-Beidai He
Heng Shan
Peking Man Site at Zhoukoudian
Dalian Seashore-Lüshunkou
Wutai Shan
Western Qing Tombs
Tianjin
BO HAI
Yinchuan
Shijiazhuang
Beach of Jiaodong Peninsula
Western Xia Imperial Tombs
Taiyuan
Cangyan Shan
Xining
Ancient City of Pingyao
Ji'nan
Lao Shan
Tai Shan
YELLOW SEA
Lanzhou
Hukou Falls
Yin Xu in Anyang
Confucius Residence, Confucian Temple and Woods
Jinghang Yunhe
Xi'an
Hua Shan
Zhengzhou
Yuntai Shan
Maiji Shan
Mausoleum of the First Qin Emperor and the Terracotta Army
Longmen Grottoes
Song Shan
Jiuzhaigou
Huanglong
Shugang Shouxihu
Wudang Shan
Jigong Shan
Hefei
Nanjing
Suzhou Classical Gardens
Panda Sanctuaries
Jianmen Shudao
Xiaoling Tomb
Tai Hu
Shanghai
Three Gorges of Chanjiang
Dahong Shan
Xianling Tomb
Wuhan
Tianzhu Shan
Huang Shan
Hangzhou
Putuo Shan
Huanglong
Chengdu
Chang Jiang
Jiuhua Shan
Xi Hu
Jinyun Shan
Dong Hu
Wannan Ancient Villages
Dazu Rock Carvings
Lushan National Park
Fuchun Jiang-Xin'an Jiang
Chongqing
Wulong Karst
Wulingyuan
Dongting Hu
Yueyanglou
Poyang Hu
EAST CHINA SEA
Dongting Hu
Sanqing Shan
Yandang Shan
Nanchang
Shunan Zhuhai
Changsha
Longhu Shan
Wuyi Shan
Taimu Shan
Chiwei Yu
Zhijindong
Guiyang
Wuyang He
Heng Shan
Diaoyu Dao
Hongfeng Hu
Jinggang Shan
Fuzhou
Libo Karst
Qingyuan Shan
Huangguoshu Waterfall
Fujian Tulou
Taiwan Strait
Taipei
Guilin Li Jiang
Danxia Shan
Gulang Yu-Wanshi Shan
Shilin Karst
Zhaoqing Xing Hu
Taiwan Dao
Guiping Xi Shan
Guangzhou
Nanning
Xiqiao Shan
Hong Kong
Hua Shan
Kaiping Diaolou and Village
Macau
Historic Centre of Macau
Dongsha Qundao
SOUTH CHINA SEA
Haikou
Hainan Dao
Tropical Seashore of Sanya
LEGEND
World Heritage Site
National Key Scenic Area
Nanning
Guangzhou
Taiwan Dao
Hong Kong
Macau
Haikou
Dongsha Qundao
Hainan Dao
Xisha Qundao
Huangyan Dao
Zhongsha Qundao
SOUTH CHINA SEA
Nansha Qundao
Zengmu Ansha
South China Sea Islands

Rare Animals in China

In 2004, the Chinese government announced the Law of the People's Republic of China on the Protection of Wildlife, classifying wildlife into the first-grade and second-grade state protection categories.

In China's *List of Wildlife under Special State Protection* announced in 1988, animals under the first-grade state protection include giant pandas, golden monkeys, long-armed apes, red-crowned cranes, baiji dolphins, Tibetan antelopes, Chinese giant salamanders, snow leopards, Amur tigers, South China tigers, and Chinese alligators, amounting to over 90 kinds of animals. Among them, giant pandas are the most well-known and popular rare species. It even became the logo of the World Wildlife Fund (WWF).

Rare Animal Species in China

China is home to one of the world's most kinds of wild animals. However, due to reasons like long-term pollution, overhunting, and natural disasters, the number of rare animals has dropped drasticly. Some animals including Amur tigers, giant pandas, golden monkeys, Tibetan antelopes, baiji dolphins, and Chinese alligators are facing extinction. They are in need of extra protection.

Tibetan Antelope

Baiji Dolphin

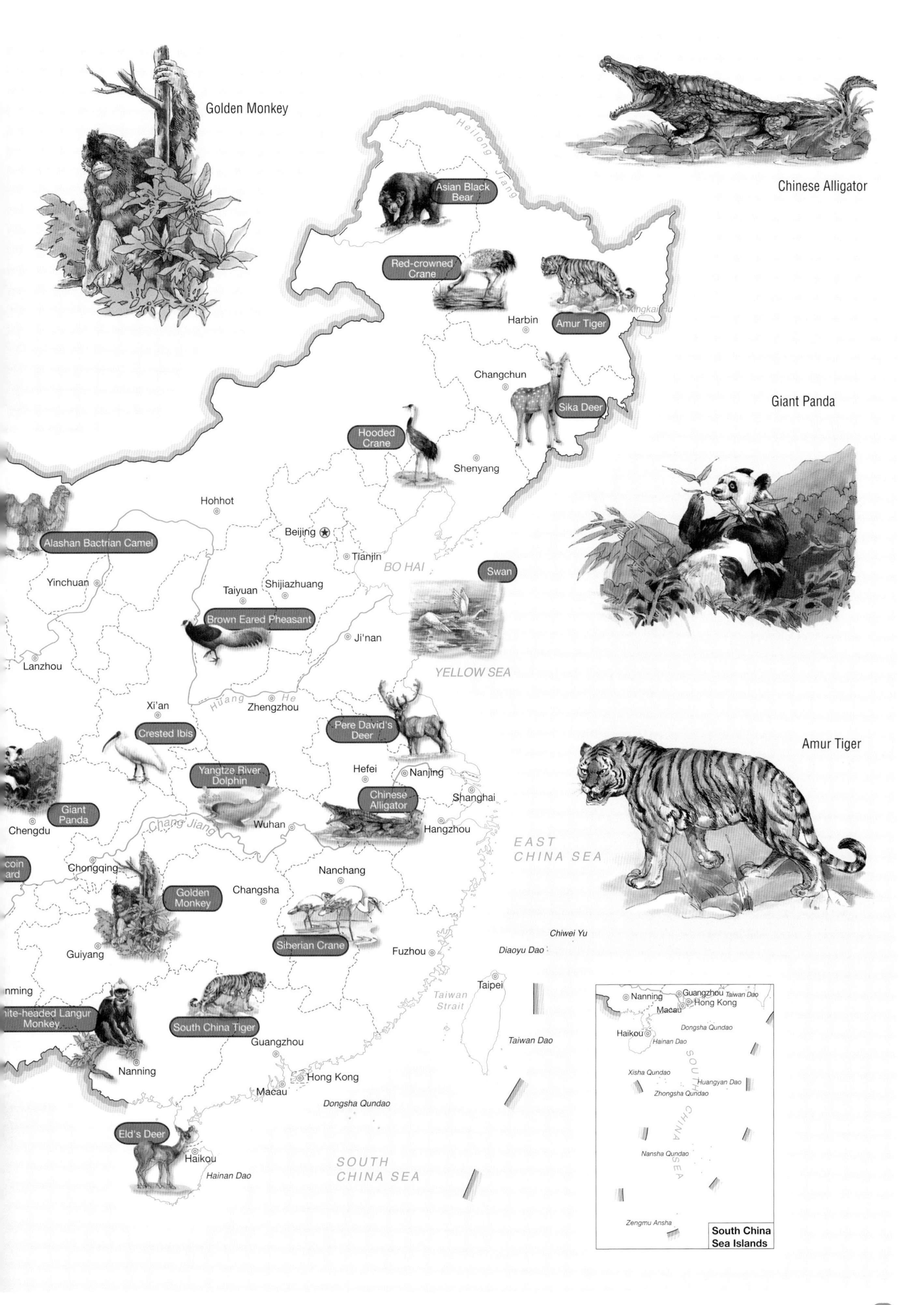
Golden Monkey
Chinese Alligator
Giant Panda
Amur Tiger
Heilong Jiang
Asian Black Bear
Red-crowned Crane
Amur Tiger
Harbin
Xingkai Hu
Changchun
Sika Deer
Hooded Crane
Shenyang
Hohhot
Alashan Bactrian Camel
Beijing
Tianjin
BO HAI
Swan
Yinchuan
Taiyuan
Shijiazhuang
Brown Eared Pheasant
Ji'nan
Lanzhou
YELLOW SEA
Xi'an
Huang He
Zhengzhou
Crested Ibis
Pere David's Deer
Yangtze River Dolphin
Hefei
Nanjing
Chinese Alligator
Shanghai
Giant Panda
Chengdu
Chang Jiang
Wuhan
Hangzhou
EAST CHINA SEA
Chongqing
Golden Monkey
Changsha
Nanchang
Siberian Crane
Fuzhou
Chiwei Yu
Diaoyu Dao
Guiyang
Taipei
Taiwan Strait
White-headed Langur Monkey
South China Tiger
Guangzhou
Taiwan Dao
Nanning
Hong Kong
Macau
Dongsha Qundao
Eld's Deer
Haikou
Hainan Dao
SOUTH CHINA SEA
Nanning
Guangzhou
Taiwan Dao
Hong Kong
Macau
Haikou
Dongsha Qundao
Hainan Dao
Xisha Qundao
Huangyan Dao
Zhongsha Qundao
SOUTH CHINA SEA
Nansha Qundao
Zengmu Ansha
South China Sea Islands

Local Specialities in China

Intangible Cultural Heritage

Among the various kinds of specialities in China, some are regarded as Intangible Cultural Heritage. Most of them are oral traditions of ethnic groups, traditional performing arts, folk events and rituals, knowledge and practice of ritual traditions, and traditional handicrafts and skills. China holds many intangible cultural heritage elements, including Kun Qu opera, Guqin and its music, Uygur Muqam of Xinjiang, Urtiin Duu, traditional folk long song, sericulture and silk craftsmanship of China, Nanyin, craftsmanship of Nanjing Yunjin brocade, traditional handicrafts of making Xuan paper, grand song of the Dong ethnic group, Yueju opera, Gesar epic tradition, traditional firing technology of Longquan celadon, Regong arts, Tibetan opera, Manas, Mongolian art of singing (Khoomei), Hua'er, Xi'an wind and percussion ensemble, Farmers' Dance of China's Chaoxian ethnic group, Chinese calligraphy, art of Chinese seal engraving, Chinese paper-cut, China engraved block printing technique, wooden movable-type printing of China, Dragon Boat Festival, Matsu belief and customs, Peking opera, acupuncture and moxibustion of traditional Chinese medicine, and Chinese shadow puppetry.

Unique Local Treasures in China

The widely diversed Chinese local specialities consist of natural products, elaborate handicrafts, and mouth-watery delicacies. They are world-famous for their wide variety and exceptional quality.

Guizhou Maotai

Three Treasures of Northern China – Ginseng, Mink Skin, and Deer Antler

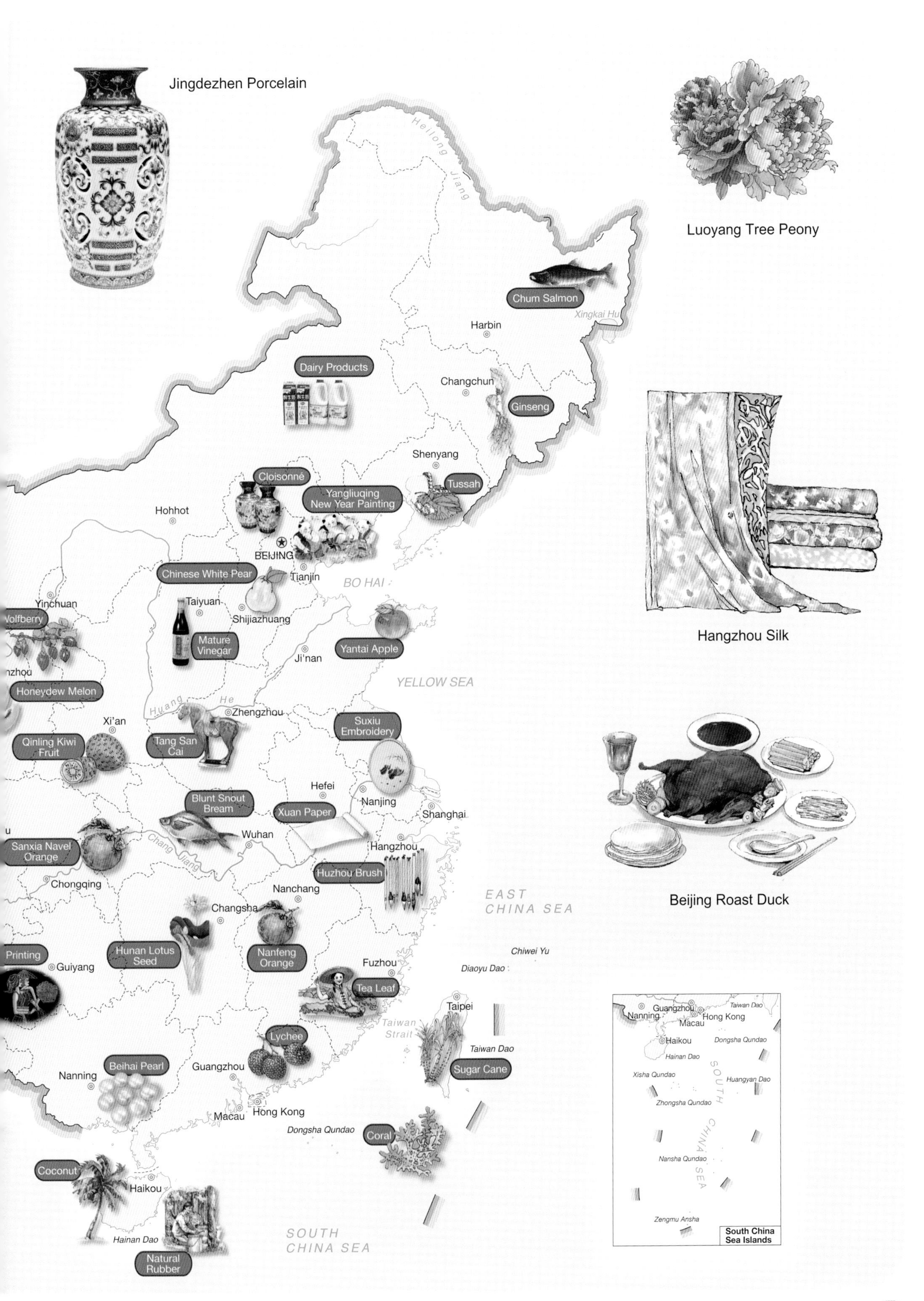

Jingdezhen Porcelain
Luoyang Tree Peony
Hangzhou Silk
Beijing Roast Duck
Heilong Jiang
Chum Salmon
Xingkai Hu
Harbin
Dairy Products
Changchun
Ginseng
Shenyang
Tussah
Cloisonné
Yangliuqing New Year Painting
Hohhot
BEIJING
Tianjin
Chinese White Pear
BO HAI
Yinchuan
Taiyuan
Shijiazhuang
Mature Vinegar
Yantai Apple
Ji'nan
Honeydew Melon
YELLOW SEA
Huang He
Zhengzhou
Xi'an
Qinling Kiwi Fruit
Tang San Cai
Suxiu Embroidery
Hefei
Nanjing
Blunt Snout Bream
Xuan Paper
Shanghai
Wuhan
Sanxia Navel Orange
Chang Jiang
Hangzhou
Huzhou Brush
Chongqing
Nanchang
EAST CHINA SEA
Changsha
Printing
Guiyang
Hunan Lotus Seed
Nanfeng Orange
Fuzhou
Chiwei Yu
Diaoyu Dao
Tea Leaf
Taipei
Taiwan Strait
Lychee
Taiwan Dao
Sugar Cane
Beihai Pearl
Nanning
Guangzhou
Macau
Hong Kong
Dongsha Qundao
Coral
Coconut
Haikou
Hainan Dao
Natural Rubber
SOUTH CHINA SEA
Taiwan Dao
Dongsha Qundao
Hainan Dao
Xisha Qundao
Huangyan Dao
Zhongsha Qundao
SOUTH CHINA SEA
Nansha Qundao
Zengmu Ansha
South China Sea Islands

Beijing

Beijing, a city of more than 3,000 years and a capital of more than 850 years, is one of China's 'Four Ancient Capital Cities.'

Beijing was first recorded as Ji. Ji was a vassal state with its capital in around the present-day Guang'anmen in Beijing. The king, surnamed 'Yi' (some say 'Qi') was the offspring of the ancient ruler Yao. In 1046 B.C., Emperor Wu of Zhou ended the Shang Dynasty and built the Western Zhou Dynasty. He gave the state of Yan (present-day Fangshan district in southwest Beijing) to his brother Zhao Gong Shi. Yan defeated the state of Ji in 700 B.C. and moved its capital to Ji. Since then, Beijing has become 'Yanjing,' the capital of Yan. During the reign of Yongle of Ming, the Emperor moved the capital from Nanjing (Nan means south) to Beijing (Bei means north). As Nanjing was still kept as the alternate capital, the name 'Beijing' was given to the capital in the north.

① Badaling Great Wall

The Ming Great Wall in Beijing is relatively well-conserved, among which are the most famous attractions: Badaling Great Wall, Mutianyu Great Wall, and Simatai Great Wall.

② Peking Opera

Combining singing, martial arts, dancing, acrobatic arts, and music, Peking Opera is a unique genre in opera and an essence of Chinese culture.

③ Siheyuan

Siheyuan is a traditional type of house in old-town Beijing. It comprises four buildings (the northern as the parlour and the owner's bedroom, eastern and western as guestrooms, and southern as servants' room) surrounding a central courtyard. The closed layout provides a tranquil, safe environment and protects residents from wind and sand.

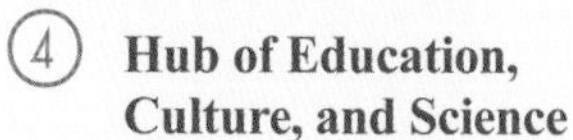

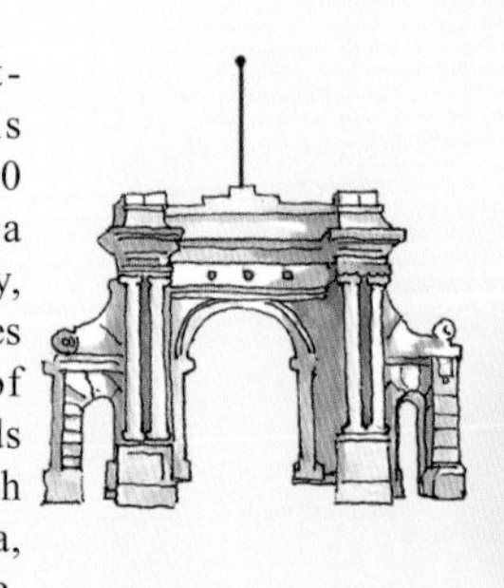

④ Hub of Education, Culture, and Science

In Beijing, there are first-rate educational institutions and cultural facilities: over 70 schools including Tsinghua University, Peking University, more than 500 research institutes such as Chinese Academy of Sciences, and several hundreds of cultural and arts facilities such as National Museum of China, and Chang'an Grand Theatre. The Zhongguancun Science and Technology Park is even reputed as the 'Silicon Valley of China.'

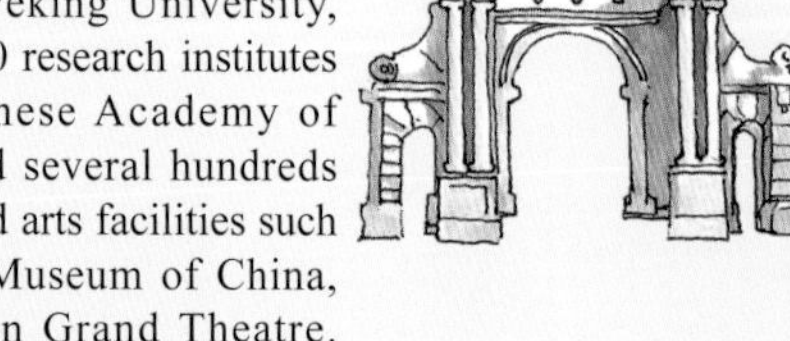

⑤ Tian'anmen Square

Tian'anmen is the world's largest urban square surrounded by the Tian'anmen gate, the Great Hall of the People, National Museum of China, Monument to the People's Heroes, Chairman Mao Memorial Hall and other buildings.

Guanting Shuiku

Hebei

Jingbai

Lingshan

Mopan Persimmon

Shidu

Juma He

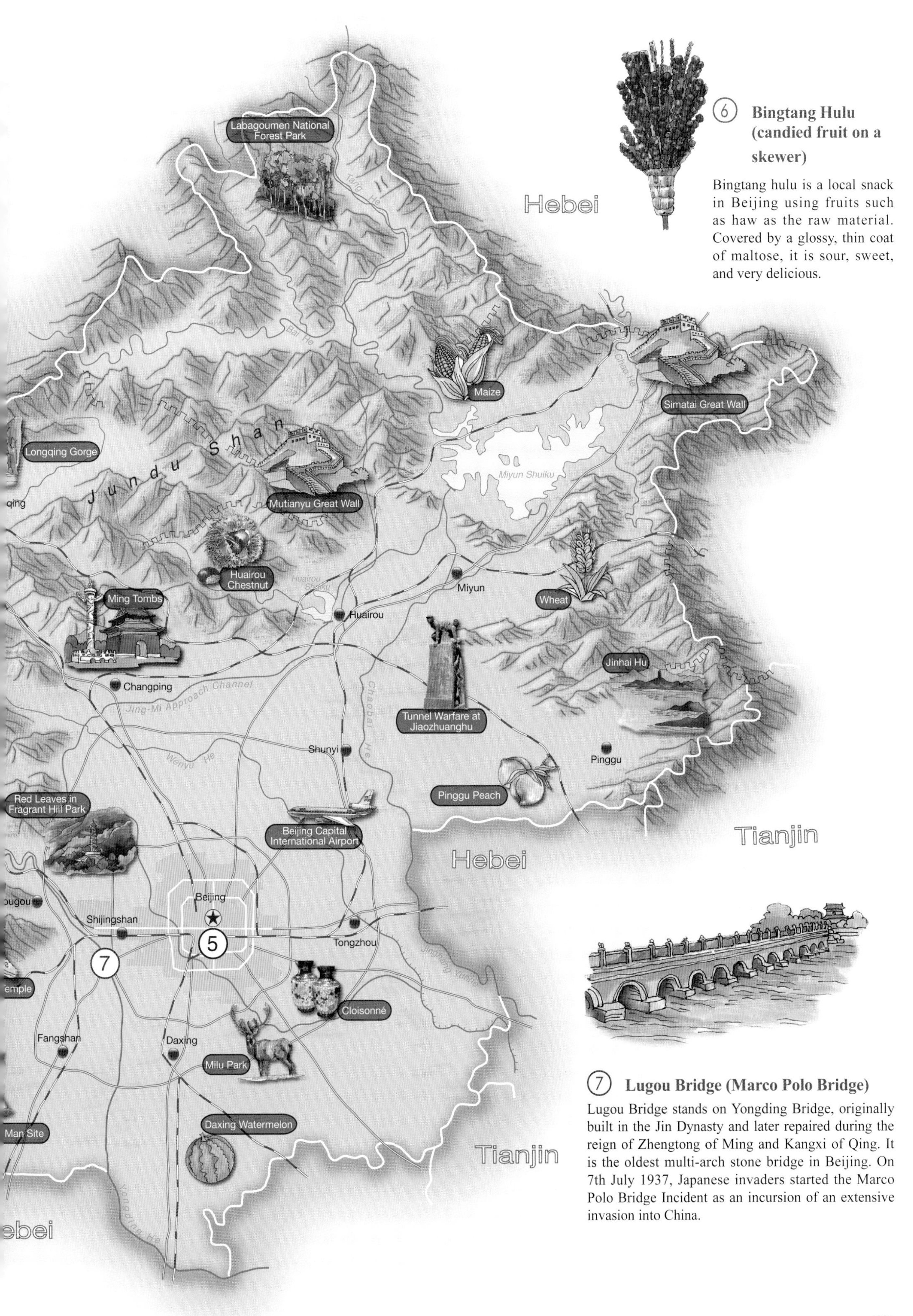

⑥ Bingtang Hulu (candied fruit on a skewer)

Bingtang hulu is a local snack in Beijing using fruits such as haw as the raw material. Covered by a glossy, thin coat of maltose, it is sour, sweet, and very delicious.

⑦ Lugou Bridge (Marco Polo Bridge)

Lugou Bridge stands on Yongding Bridge, originally built in the Jin Dynasty and later repaired during the reign of Zhengtong of Ming and Kangxi of Qing. It is the oldest multi-arch stone bridge in Beijing. On 7th July 1937, Japanese invaders started the Marco Polo Bridge Incident as an incursion of an extensive invasion into China.

Urban Area of Beijing

Beijing is the hub of Chinese traditional cultures with plentiful cultural and historical sites. It has the highest number of UNESCO World Heritage sites among all cities in the world (the Peking Man Site at Zhoukoudian, Great Wall, Imperial Palace, Summer Palace, Tiantan, and Ming Tombs).

During Liao and Jin dynasties, Beijing is the southern capital of Liao and central capital of Jin but with a slightly different scale and location from the present-day Beijing. As it became the capital of the Yuan, Ming, and Qing dynasties, it began to look like the present-day Beijing. The basic townscape was built by Emperor Chengzu of Ming, such as the Imperial Palace (also known as the Forbidden City) at the centre of Beijing, and the core urban area of the city within the Erhuan Lu. Ancient names of many gates and buildings are still in use, such as Deshengmen, Xuanwumen, Zhengyangmen, and Andingmen.

① Wangfujing Dajie

Wangfujing Dajie is one of the most famous commercial streets in Beijing and the first choice for shoppers.

② Imperial Palace

Also known as the Forbidden City, the Imperial Palace is the residence and working place of Ming and Qing emperors as well as the world's largest palace complex.

③ National Stadium and National Aquatics Centre

National Stadium ('Bird's Nest') and National Aquatics Centre ('Water Cube') are two of the major sports centres of the 29th Olympic Games held in 2008 and landmarks of Beijing.

④ Tiantan

Tiantan is where emperors of Ming and Qing dynasties worship and pray for good harvests.

⑤ National Grand Theatre of China

Located in the west of the Great Hall of the People, the National Grand Theatre of China is the best performing arts centre in China.

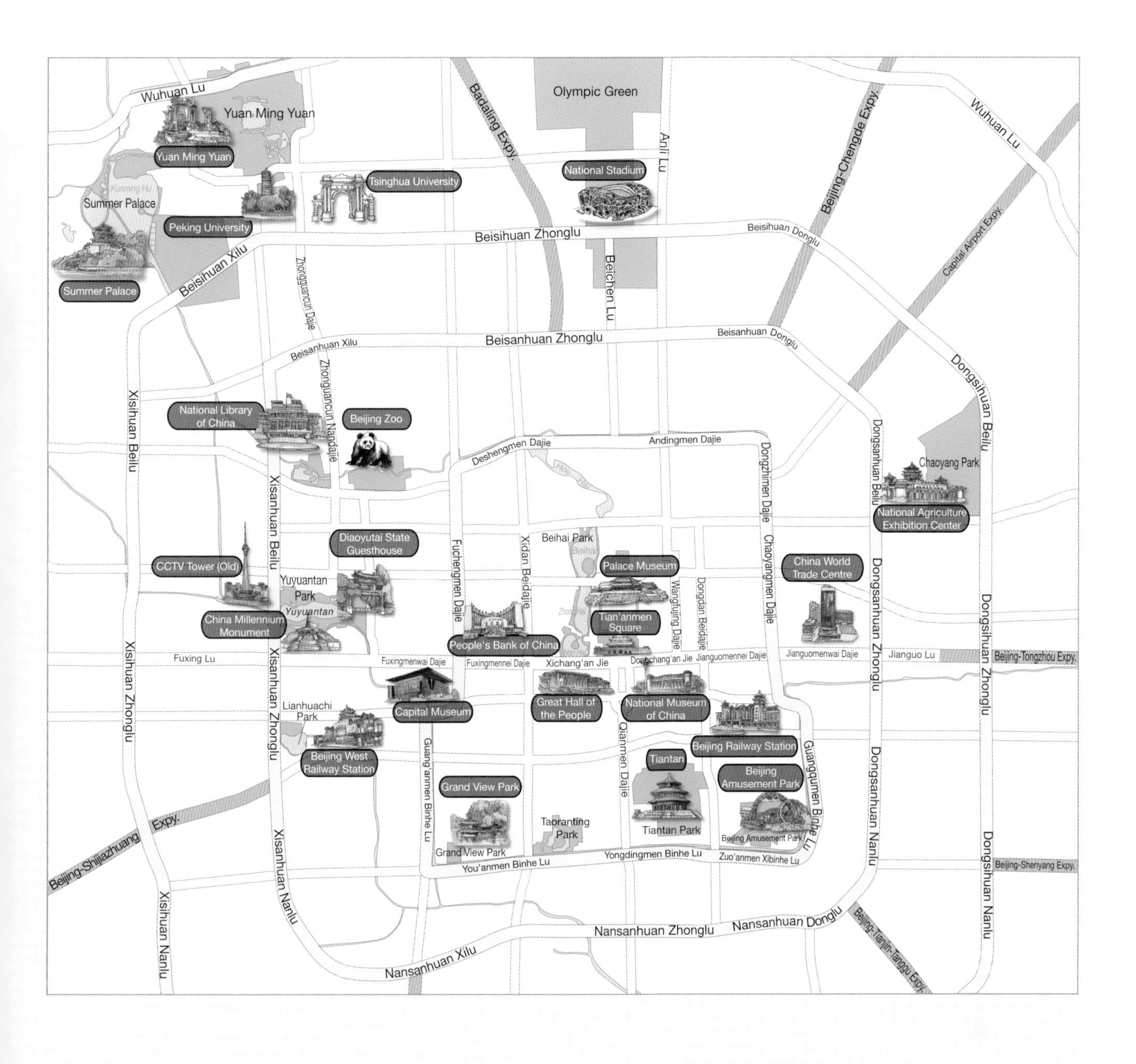

⑥ Chang'an Jie

Chang'an Jie is known as the 'Most Important Street in China.' Both sides of the street are lined by many buildings of government offices and large infrastructures such as the Tian'anmen Square, the Great Hall of the People, Zhongnanhai (the headquarters of the Chinese Communist Party), the Imperial Palace, and the National Museum of China.

① **Pan Shan**

Located in Jixian of Tianjin, Pan Shan is known as the 'Most Famous Mountain in Jingdong' and the only national-level major scenic site in Tianjin.

Tianjin

Tianjin was officially built on the 21st day of the 11th month in the Chinese calendar (23rd December 1404). It was the only ancient city in China with recorded date of construction.

The name Tianjin, literally meaning the 'port crossed by the Son of Heaven (the imperial title of a Chinese emperor),' was given by Emperor Chengzu of Ming. Later Tianjin gradually became an essential coastal city for protecting Beijing. As the Second Opium War broke out in 1860, the Anglo-French Army landed on Dagu, Tianjin, then captured Beijing and set fire to Yuan Ming Yuan. In 1900, the eight-nation allied army also landed on Tianjin and captured the city of Beijing.

In early years of the Republic of China, many retired politicians and veterans from the Qing Dynasty fled to foreign concession in Tianjin for shelter. In February of 1925, the last emperor of the Qing Dynasty, Emperor Puyi, entered Zhangyuan and Jingyuan, both in concession in Tianjin. After Japanese captured Jilin, Liaoning and Heilongjiang, Puyi arrived secretly at the northeast of China from Tianjin and became the puppet emperor of Manchukuo.

② **Yangliuqing New Year Painting**

Yangliuqing New Year Painting is a form of popular folk art named after its origin, a town called Yangliuqing. The scenes are usually about auspicious moments and celebrations.

③ **Three Unique Snacks of Tianjin**

The renowned 'Three Unique Snacks of Tianjin' are Goubuli steamed stuffed buns, Shibajie fried dough twist, and Erduoyan fried rice cakes.

④ **Hai He Scenic Area**

Hai He flows through the urban centre of Tianjin like a long scroll of landscape painting. The riversides are ideal for sight-seeing and relaxation.

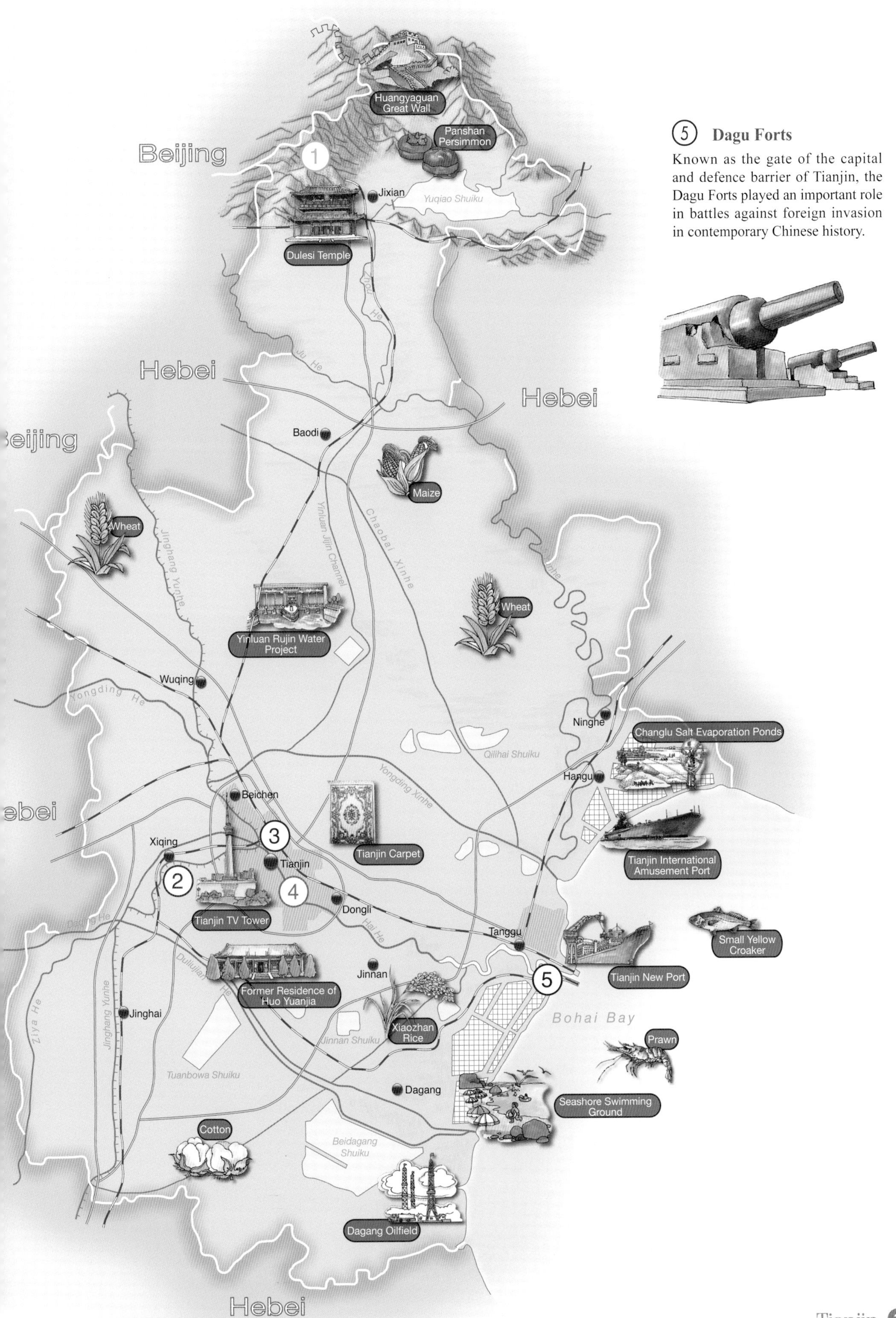

⑤ Dagu Forts

Known as the gate of the capital and defence barrier of Tianjin, the Dagu Forts played an important role in battles against foreign invasion in contemporary Chinese history.

① **Zhaozhou Bridge**

Zhaozhou Bridge is no in southern Zhaoxia and the world's oldes extant stone arch bridg with over 1,400 years c history.

② **Western Qing Tombs**

Western Qing Tombs are at the feet of Yongning in mid-western Hebei. They are a UNESCO historical site and one of the Imperial Tombs of the Qing Dynasty. Another famous imperial tomb, Eastern Qing Tombs, is also in Hebei.

Hebei Province

Hebei, meaning 'north of river,' is given the name as the province is in the north of Huang He. As most of its land was within 'Jizhou,' one of the legendary 'Nine Provinces,' the province is abbreviated as 'Ji.'

Yi Shui and Assassination Attempt

During the late period of the Warring States period, the State of Qin grew strong and kept seizing surrounding states. The crown prince of the State of Yan recruited a heroic man named Jing Ke to kill Ying Zheng, the First Emperor of Qin. The crown prince prepared a farewell meal for him beside Yi Shui. Jing Ke sang solemnly by the river to express his determination about the assassination. Yi Shui, originated in mid-western Hebei, was in the State of Yan.

Idiom: Han Dan Xue Bu

This is a fable from *Zhuang Zi*: A man from the State of Chu heard about the elegant walking style of the people in Handan. He went a long way to get to the State of Zhao to learn from them, but he failed and even forgot his own way of walking. He finally had to crawl back home. The phrase has been used to describe people who failed in imitating others and finally forgot their own skills. Handan was then the capital of Zhao and now a city in Hebei.

③ **Shanhaiguan**

Shanhaiguan is the easternmost starting point of the Ming Great Wall. Well-known as the 'First Pass in China,' it was an important gate to the transport network within and out of the pass.

④ **Mountain Resort**

The Mountain Resort in Chengde was once the summer resort for Qing emperors. Now it is a famous classical imperial garden in China.

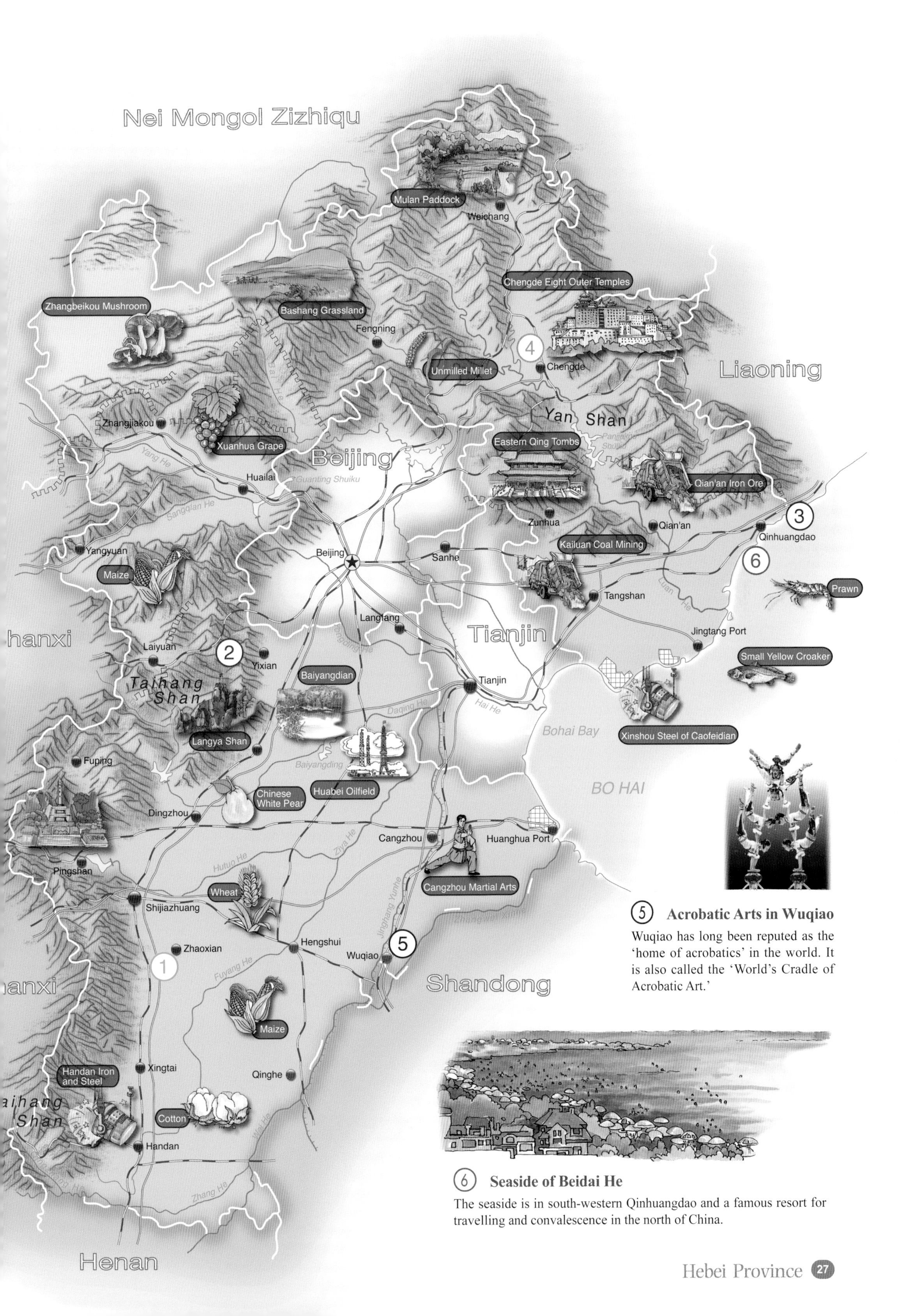

⑤ Acrobatic Arts in Wuqiao

Wuqiao has long been reputed as the 'home of acrobatics' in the world. It is also called the 'World's Cradle of Acrobatic Art.'

⑥ Seaside of Beidai He

The seaside is in south-western Qinhuangdao and a famous resort for travelling and convalescence in the north of China.

① **Ancient City of Ping Yao**

Ping Yao is the best preserved historic city from the Ming and Qing dynasties. It was included in the *UNESCO World Heritage List* in 1997.

Shanxi Province

The name Shanxi literally means 'west of mountains' as the region is located in the west of Taihang Shan. In the Zhou Dynasty, Shanxi belonged to the State of Tang. Later, the state was renamed as Jin, which has become the abbreviation of Shanxi. The Tang Dynasty (618—907 A.D.) was named after Li Yuan, the first emperor of the dynasty, who inherited the name of the 'Duke of the State of Tang' before he gathered power in Taiyuan.

Cold Food Festival and Shanxi

During the Spring and Autumn period, an aristocrat of the State of Jin named Jie Zitui fled to other states with the Crown Prince of Jin, Chong'er, for over ten years. During the exile, Jie cut a piece of flesh off his thigh to save Chong'er who was suffering from hunger and coldness. After Chong'er ascended the throne as Wengong of Jin, Jie refused to compete for honour, so he lived in seclusion with his mother in Mian Shan. To force him into officialdom, Wengong ordered to set the mountain on fire but it led to the death of Jie and his mother. Wengong was deeply in grief. He then renamed the region as Jiexiu and the mountains as Jie Shan. He also named the day before the Qingming Festival as the Cold Food Festival so that no fire was allowed and every household could only eat cold food. Since then it has become an important event in China.

② **Yungang Grottoes**

Located in the countryside in the west of Datong, the Yungang Grottoes have more than 1,500 years of history and over 50,000 Buddhist statues. They are one of China's three great Buddhist grottoes (the other two are Longmen Grottoes in Luoyang and Mogao Caves in Dunhuang) and a UNESCO World Heritage site.

③ **Hanging Temple—the most popular site in Heng Shan**

Clinging to a sheer cliff-face, the Hanging Temple is awe-inspiring because of the delicate structure and precarious location.

④ **Hukou Waterfall of Huang He**

Hukou Waterfall is a national-level major scenic site situated in Hukou in the southeast of Jixian. It only comes second to the Huangguoshu Waterfall in Guizhou.

⑤ **Wutai Shan**

Wutai Shan is one of China's Four Sacred Mountains of Buddhism. The cool, pleasant climate in summer and the beautiful scenery attract many tourists to come and escape from the heat.

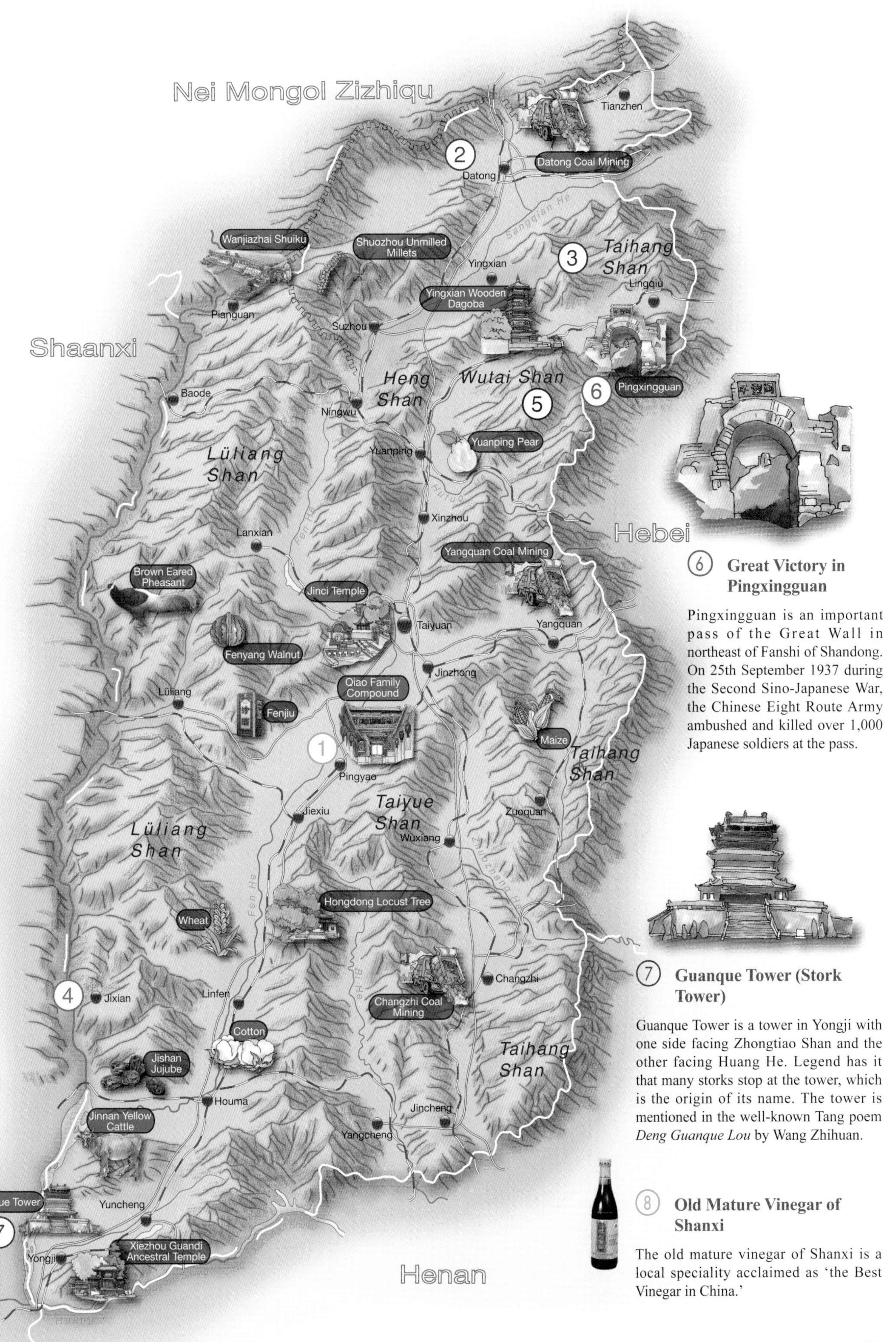

⑥ Great Victory in Pingxingguan

Pingxingguan is an important pass of the Great Wall in northeast of Fanshi of Shandong. On 25th September 1937 during the Second Sino-Japanese War, the Chinese Eight Route Army ambushed and killed over 1,000 Japanese soldiers at the pass.

⑦ Guanque Tower (Stork Tower)

Guanque Tower is a tower in Yongji with one side facing Zhongtiao Shan and the other facing Huang He. Legend has it that many storks stop at the tower, which is the origin of its name. The tower is mentioned in the well-known Tang poem *Deng Guanque Lou* by Wang Zhihuan.

⑧ Old Mature Vinegar of Shanxi

The old mature vinegar of Shanxi is a local speciality acclaimed as 'the Best Vinegar in China.'

Nei Mongol Zizhiqu

In the Qing Dynasty, the place where Monan Mongols live was called Inner Mongolia while Mobei Mongols live in Outer Mongolia. Since then the name Inner Mongolia or Nei Mongol has been used.

Yin Shan is in Inner Mongolia and seen as the boundary between Chinese agricultural culture and the nomadic culture. *Chi Le Ge*, a famous folk song from the South and North Dynasties, depicts the expansive grassland at the feet of Yin Shan. It is also mentioned in the poem *Chu Sai* by the Tang poet Wang Changling.

Idiom: Zhao Jun Chu Sai

The Huns are a minor ethnicity group living in Mongolian grasslands in northern China. During the reign of Emperor Yuan of the Han Dynasty, the Hun chief requested for marriage alliance to Han. A court maid named Wang Zhaojun took the initiative and became the chief's wife, achieving peace between Han and Hun for more than a century. Her story was one of the inspirations of *Yonghuai Guji* written by the Tang poet Du Fu.

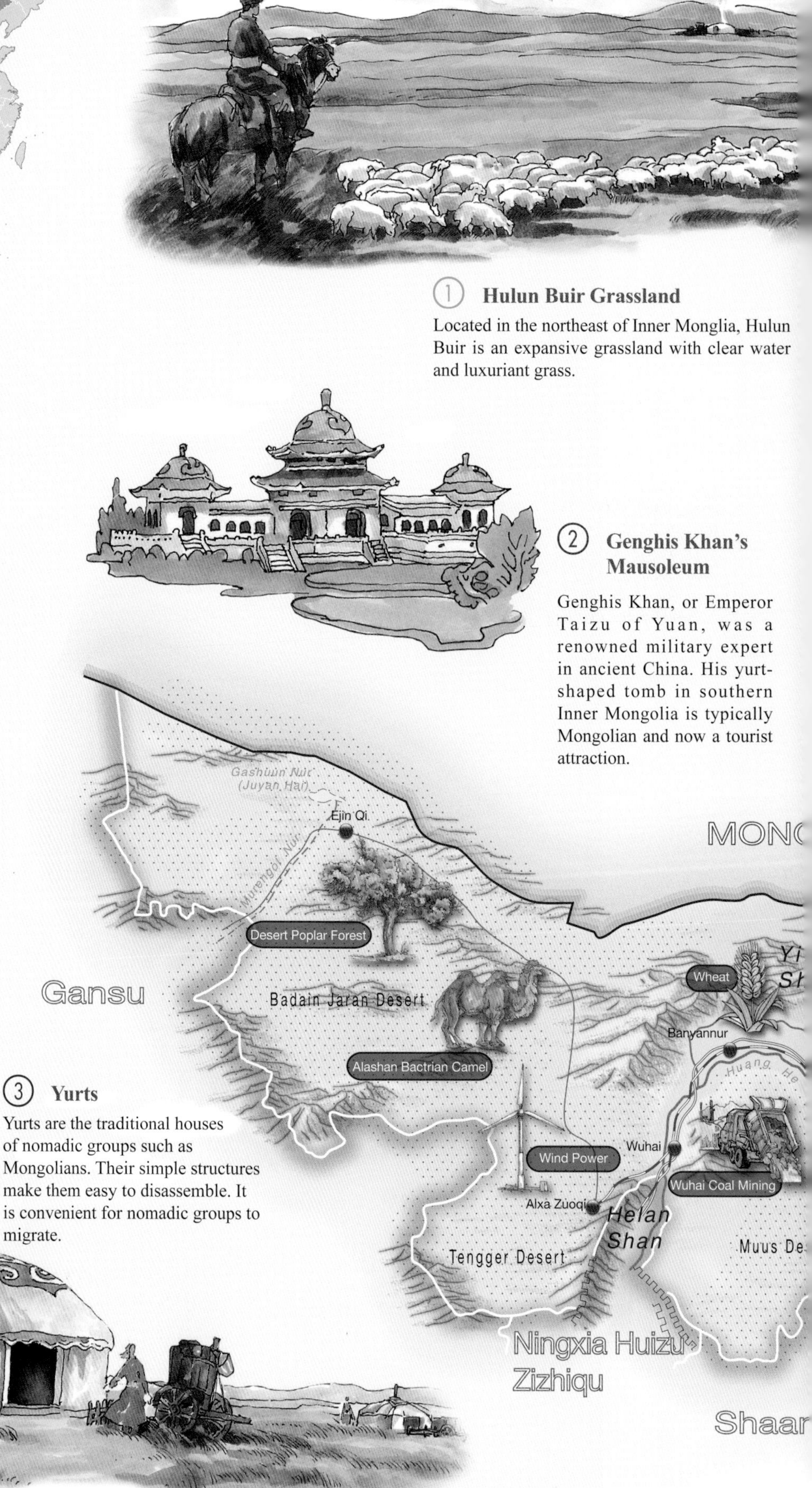

① **Hulun Buir Grassland**

Located in the northeast of Inner Monglia, Hulun Buir is an expansive grassland with clear water and luxuriant grass.

② **Genghis Khan's Mausoleum**

Genghis Khan, or Emperor Taizu of Yuan, was a renowned military expert in ancient China. His yurt-shaped tomb in southern Inner Mongolia is typically Mongolian and now a tourist attraction.

③ **Yurts**

Yurts are the traditional houses of nomadic groups such as Mongolians. Their simple structures make them easy to disassemble. It is convenient for nomadic groups to migrate.

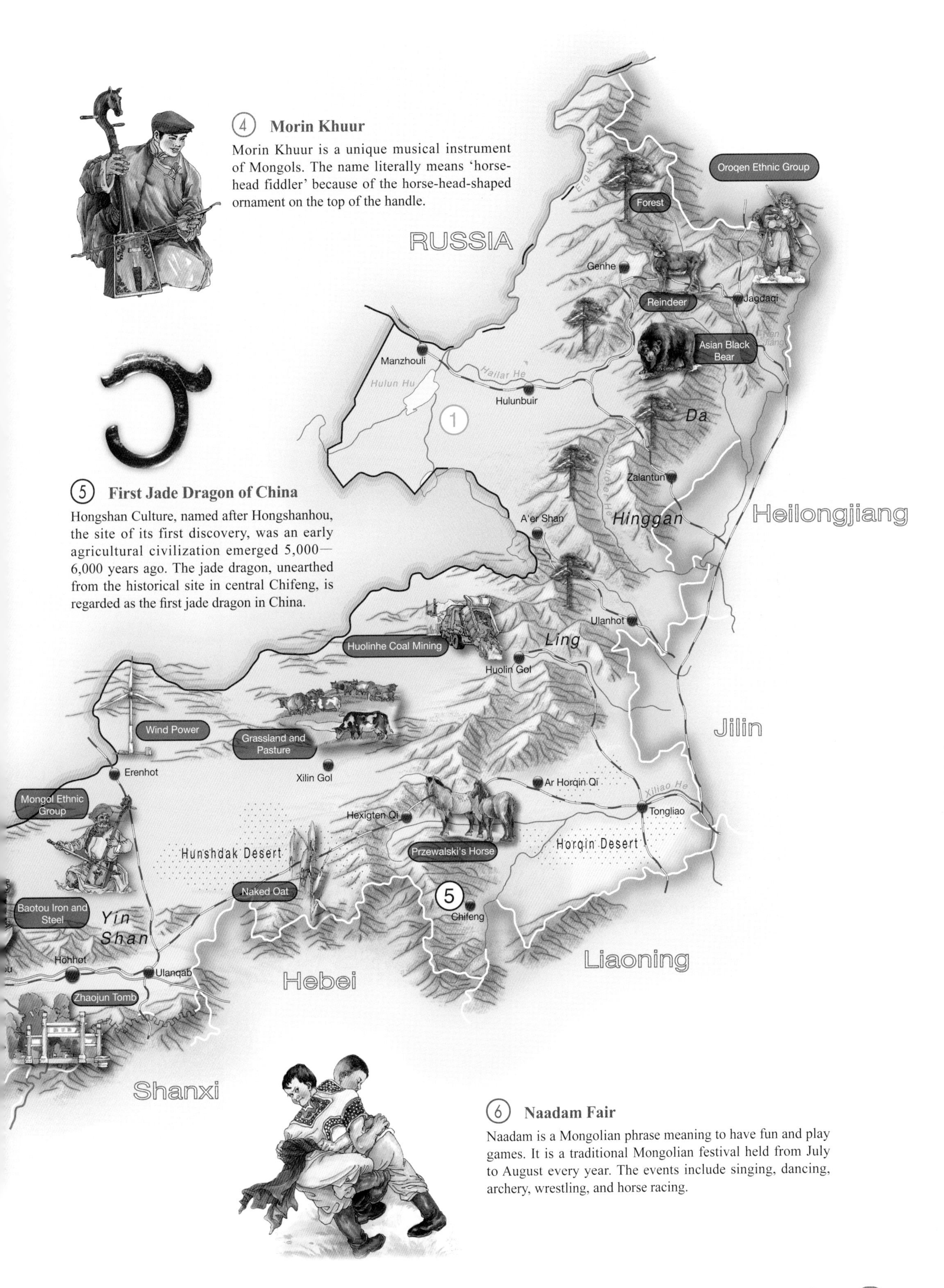

④ Morin Khuur

Morin Khuur is a unique musical instrument of Mongols. The name literally means 'horse-head fiddler' because of the horse-head-shaped ornament on the top of the handle.

⑤ First Jade Dragon of China

Hongshan Culture, named after Hongshanhou, the site of its first discovery, was an early agricultural civilization emerged 5,000—6,000 years ago. The jade dragon, unearthed from the historical site in central Chifeng, is regarded as the first jade dragon in China.

⑥ Naadam Fair

Naadam is a Mongolian phrase meaning to have fun and play games. It is a traditional Mongolian festival held from July to August every year. The events include singing, dancing, archery, wrestling, and horse racing.

① **9.18 Historical Museum**

On 18th September 1931, Japanese army in north-eastern China ambushed Shenyang, and soon captured Liaoning, Jilin, and Heilongjiang as a pretext of its invasion into China. The day was marked as the '918 Incident' and the museum built in Shenyang was to remind people of the humiliation that China has borne.

Liaoning Province

The Liao He region is called by the Manchurians 'Fengtian Province,' where they rose to power and founded the Qing Dynasty. 'Fengtian' means 'receiving the mandate from Heaven to rule the country.' In 1920s, it was named after the Liao He as Liaoning Province. 'Liaoning' denotes a blessing that the Liao He region will be peaceful forever.

Birthplace of the Qing Dynasty

In the 1st month of 1616, Aisin Gioro Nurhaci, the leader of the Jurchen (later as Manchu) people, called himself Khan of the Jin Dynasty (or Hou Jin Dynasty by later generations) in Hetuala, present-day around Xinbin. Later the Jin Dynasty grew stronger and finally united China to build the last dynasty, the Qing Dynasty.

First Sino-Japanese War (1894—1895)

Liaoning was a major battlefield of the war and where the Japanese army conducted the appalling Lüshun Massacre. After the Qing government lost, Japan forced it to offer several parts of its land, including the Liaoning Peninsula, and 200 million *liang*. Later, it gave Japan another 30 million *liang* under the influence of other foreign forces as a 'ransom' for taking back Liaoning.

② **Fossil of Chinese Dragon Bird**

Found in Chaoyang in western Liaoning, the fossil of Chinese Dragon Bird showed that the ancestor of birds was small dinosaurs. Chaoyang has become an internationally well-known treasure of bird fossils.

③ **Shenyang Imperial Palace**

Shenyang Imperial Palace was the palace of early Qing dynasty, only second to the Forbidden City in Beijing among extant palaces in China.

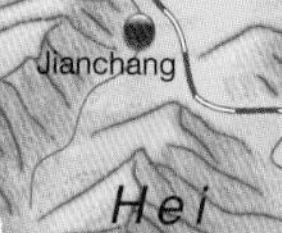

④ **Dalian**

Situated in the most southern corner of Liaodong Peninsula, Dalian has beautiful scenery and a pleasant climate. It is not only an important port in northern China, but also a famous seaside destination to escape from heat.

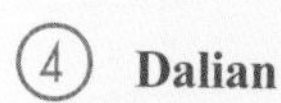

⑤ Manchu

Manchu is the second largest ethnic group in China. Manchu people mainly live in Liaoning and their traditional costume include qipao, magua, and xuezi.

⑥ Dandong (China-DPRK Friendship Bridge)

As the US army landed on the Korean Peninsula when the Korean War broke out in 1950, the Chinese People's Volunteer Army immediately crossed Yalu Jiang in Dandong to fight in Korea, finally forcing the US to sign a ceasefire agreement in 1953.

Jilin Province

In 1653, the Qing government set up the Ningguta area in present-day Jilin. That was the start of Jilin's development as a province. Later it was renamed as Ningguta General. In 1673, the city Jilin was built and was then called 'Jilinwula' (meaning 'along the river'). In 1757, the Ningguta General was changed into Jilin General. Since then the meaning of 'Jilin' has expanded from a small city name to an administrative region. It became the official name of the province in 1907.

Puppet State of Manchukuo

After the 918 Incident in 1931, Japan controlled the north-eastern part of China. In March of 1932, Japan set up the puppet government of Manchukuo with Puyi as the administrator (later called 'Huangdi') and changed the name Changchun, the capital of Jilin, into Xinjing, as the capital of the puppet government. On 15th August 1945, Japan announced its unconditional surrender and the Puppet State of Manchukuo ended, but the puppet palace is still extant in Changchun.

① **Chaoxian Ethnic Group**

Chaoxian people are one of China's 56 ethnic groups and mainly live in Yanbian Chaoxian Autonomous Prefecture. The group's favourite sports are football and swings. Their changgu dance is very unique.

② **Tian Chi in Changbai Shan**

Tian Chi of Jilin is located in Baitou Feng, the highest peak of Changbai Shan. It is the boundary between China and North Korea and the most famous scenic site in Changbai Shan.

③ Wusong in Jilin

Wusong (or Shugua) is formed when fog is frozen and condensed on tree branches to form crystals. The spectacular, beautiful scenery makes it one of China's Four Natural Wonders (the others are Guilin landscapes, Stone Forest in Yunnan, and the Three Gorges in Chang Jiang).

④ Yiqi Automobiles of Changchun

The First Automobile Group Corporation (abbreviated as FAW, or 'Yi Qi'), with headquarters in Changchun, is the largest manufacturer and research and development hub of automobiles in China.

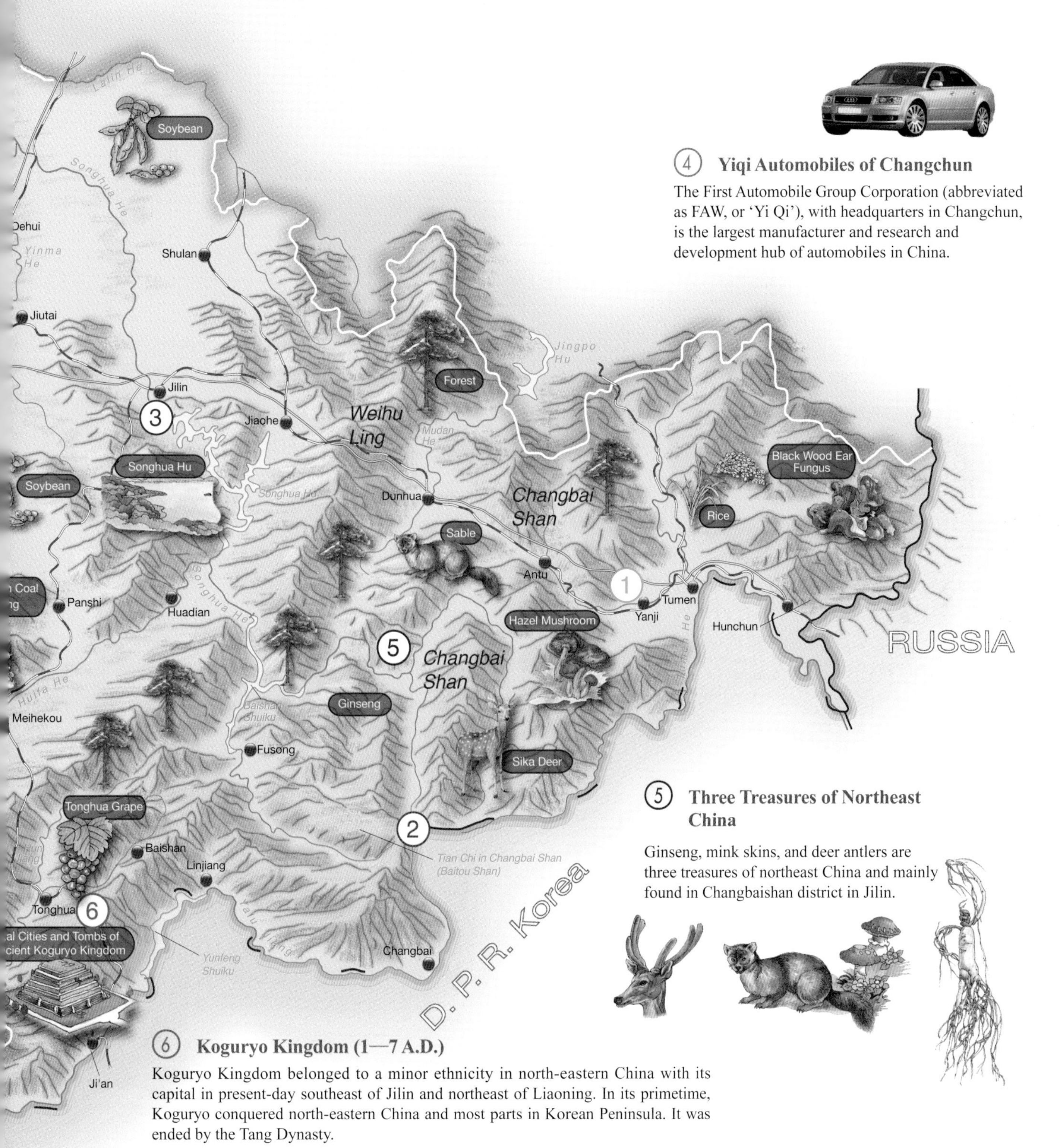

⑤ Three Treasures of Northeast China

Ginseng, mink skins, and deer antlers are three treasures of northeast China and mainly found in Changbaishan district in Jilin.

⑥ Koguryo Kingdom (1—7 A.D.)

Koguryo Kingdom belonged to a minor ethnicity in north-eastern China with its capital in present-day southeast of Jilin and northeast of Liaoning. In its primetime, Koguryo conquered north-eastern China and most parts in Korean Peninsula. It was ended by the Tang Dynasty.

Heilongjiang Province

Heilongjiang is the most northern province in China and named after a river in the region—Heilong Jiang, which marks the boundary between China and Russia. A famous Chinese wuxia fantasy novel *Ludingji* mentioned the river as 'Amur River.' The lower course was the habitat of the Heishui Mohe tribes, the ancestors of the Manchurians. In 1115 A.D., the Jurchens built the Jin Dynasty under the leadership of Wanyen Akutta and set up its capital in Harbin, the present-day capital of the province.

Wuguocheng

Wuguocheng is in north-eastern Harbin. In the Northern Song Dyansty, a well-known general Yue Fei wrote a poem *Manjianghong* mentioning the Jingkang Incident (1126—1127), in which two emperors of Northern Song were kidnapped and imprisoned by the Jurchen invaders. They were tortured to death during their imprisonment in Wuguocheng.

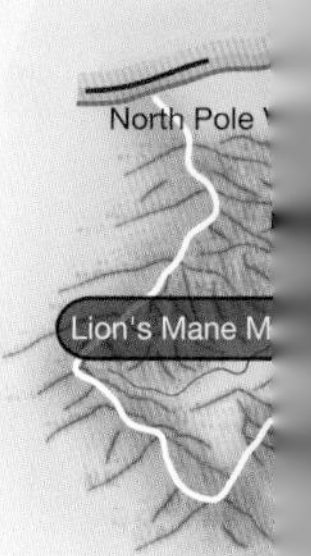

① Wudalianchi Volcanic Cluster

Wudalianchi is a major volcanic nature reserve in China and a member of Global Geoparks. It comprises 14 dormant volcanoes and 5 barrier lakes and its volcanic mineral spring is famous for treating illness.

② Yabuli Ski Resort

Yabuli Ski Resort was the venue of the 1995 Asian Winter Games. It is the perfect playground for skii-lovers.

③ Amur Tigers

Amur tigers are under first-grade state protection in China and mainly live in Da Hinggan Ling, Xiao Hinggan Ling, and Changbaishan district in north-eastern China.

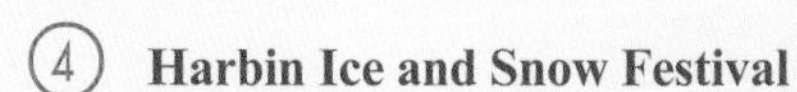

④ Harbin Ice and Snow Festival

Well-known for ice and snow art such as ice lanterns and ice and snow sculptures, the Harbin Ice and Snow Festival is one of the 4 most famous ice and snow events in the world (the others are Japan Sapporo Snow Festival, Canada Quebec Winter Carnival, and Norway Holmenkollen Ski Festival).

⑤ Zhalong Natural Reserve

The Zhalong Natural Reserve is the first large-scale natural reserve for birds living in water and home to red-crowned cranes.

⑥ Northern Pole Village

The Northern Pole Village is a small village at the northernmost border in China. Nights in winter are exceptionally long, but much shorter in summer. There is daylight on the day before and after the summer solstice for almost 24 hours each day. One might even be lucky enough to witness the splendid, multi-coloured northern lights.

Shanghai

Shanghai is abbreviated as Hu or Shen. Hu was a fishing tool made of bamboo in the Jin Dynasty and later used as a reference to the place. It is said that in the Warring States Period, Shanghai was dedicated to Huang Xie, the Lord Chunshen of the State of Chu. That is why Shanghai also goes by 'Shen.' The name Shanghai was first used in the Song Dynasty and came from a branch of Wusong Jiang called Shanghai Pu (Pu means 'small river' in the Wu dialect).

Idiom: Hua Ting He Lei

'Huating' is in Songjiang District of the present-day Shanghai. In 219 A.D. (Eastern Han), a famous general named Lu Xun captured Jingzhou tactically and forced Guan Yu, a well-known general, to flee to Maicheng and be executed. Lu Xun was honoured as the 'Marquis of Huating' for the great merit.

Lu Ji, a literati and calligrapher, was the grandchild of Lu Xun. He always visited Huating and appreciated the cries of cranes with his brothers. Later when he worked as an official in Luoyang of Henan, he was accused with a false charge and executed. Before his death, he sighed and wished he could hear the calling of cranes in Huating again.

① Yu Yuan

Yu Yuan is the largest classical garden in Shanghai. With an exquisite layout, the garden is famous for combining the art of gardening from Ming and Qing dynasties.

② Shanghai Transrapid

Shanghai Transrapid is the world's first magnetic levitation train to be put into commercial use and a symbol of Shanghai's modernization. The highest speed is up to 431 km/hour.

③ Nanjing Road

Nanjing Road is one of the busiest commercial road in Shanghai.

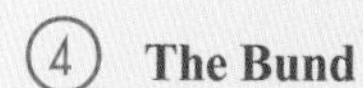

④ The Bund

The Bund is the most iconic destination in Shanghai, well-known for the beautiful western architecture of various styles. The night view at the Bund is a famous attraction and a must-see for tourists.

⑤ Pudong New District

Pudong New District reflects the modernization of Shanghai; in particular, the Oriental Pearl TV Tower by the Huangpu Jiang in Pudong is the tallest building in Asia and a landmark of Shanghai.

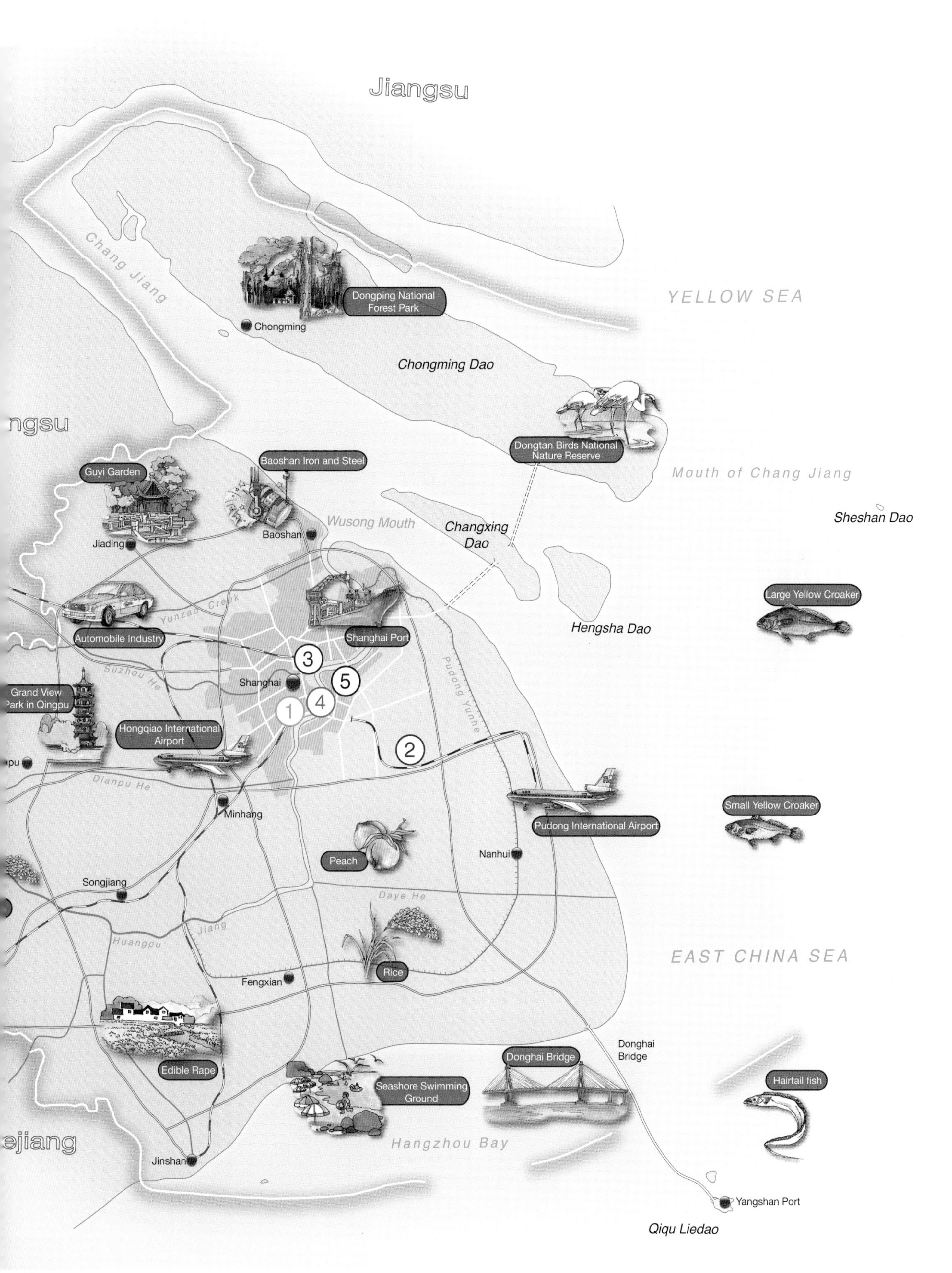

Jiangsu
ngsu
ejiang
Chang Jiang
Dongping National Forest Park
Chongming
YELLOW SEA
Chongming Dao
Dongtan Birds National Nature Reserve
Mouth of Chang Jiang
Sheshan Dao
Guyi Garden
Baoshan Iron and Steel
Jiading
Baoshan
Wusong Mouth
Changxing Dao
Large Yellow Croaker
Yunzao Creek
Automobile Industry
Shanghai Port
Hengsha Dao
Suzhou He
Shanghai
Pudong Yunhe
Grand View Park in Qingpu
Hongqiao International Airport
Dianpu He
Minhang
Small Yellow Croaker
Pudong International Airport
Nanhui
Peach
Songjiang
Daye He
Huangpu Jiang
Rice
EAST CHINA SEA
Fengxian
Edible Rape
Donghai Bridge
Donghai Bridge
Seashore Swimming Ground
Hairtail fish
Hangzhou Bay
Jinshan
Yangshan Port
Qiqu Liedao

Jiangsu Province

The name Jiangsu came from the most significant cities in the province: Jiangning (present-day Nanjing) and Suzhou.

Hanshan Temple and the City of Gusu

The city of Gusu in the Tang poem *Feng Qiao Ye Bo* by Zhang Ji is the present-day Suzhou, which was named after Gusu Shan. Also mentioned in the poem is Hanshan Temple. It receives national fame thanks to the poem and becomes a popular tourist destination.

Treaty of Nanjing

In June of 1840, the Qing army was defeated in the First Opium War. In August of 1842, the Qing government was forced to sign the Treaty of Nanjing with British representatives on the British battleship anchored in the Xiaguan Jiang in Nanjing. This was the first unequal treaty signed between China and foreign countries in contemporary Chinese history. It was under this treaty that Hong Kong Island became a British colony.

① Classical Gardens of Suzhou

Classical Gardens of Suzhou has been known as the most beautiful garden in China. There are over 100 gardens in Suzhou, among which Zhuozheng Yuan and Liu Yuan are two of China's Four Great Gardens (the other two are the Summer Palace in Beijing and the Mountain Resort in Chengde).

② Tai Hu

Tai Hu is the third largest fresh water lake in China and well-known for the pleasant scenery and the historical site of the states of Wu and Yue. It is also famous for abounding in fresh water products such as fish and shrimps.

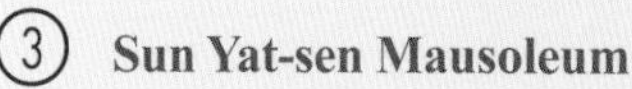

③ Sun Yat-sen Mausoleum

Lying at the feet of the southern side of Zhong Shan in Nanjing, Sun Yat-sen Mausoleum is the tomb of Dr. Sun Yat-sen, the pioneer of democratic revolution.

④ Suxiu Embroidery

Suxiu is a traditional silk embroidery handicraft with gorgeous patterns, elegant colours, and exquisite skills. It ranks the first among China's Four Great Embroideries (others are Xiangxiu, Yuexiu, and Shuxiu).

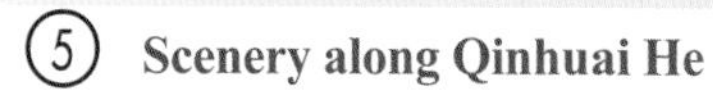

⑤ Scenery along Qinhuai He

With the Confucius Temple as the main attraction, the scenic area along the Qinhuai He is a combination of historical sites, gardens, painted boats, city streets, and folk cultures. It is a tourist destination illustrating the townscape of the ancient city of Nanjing.

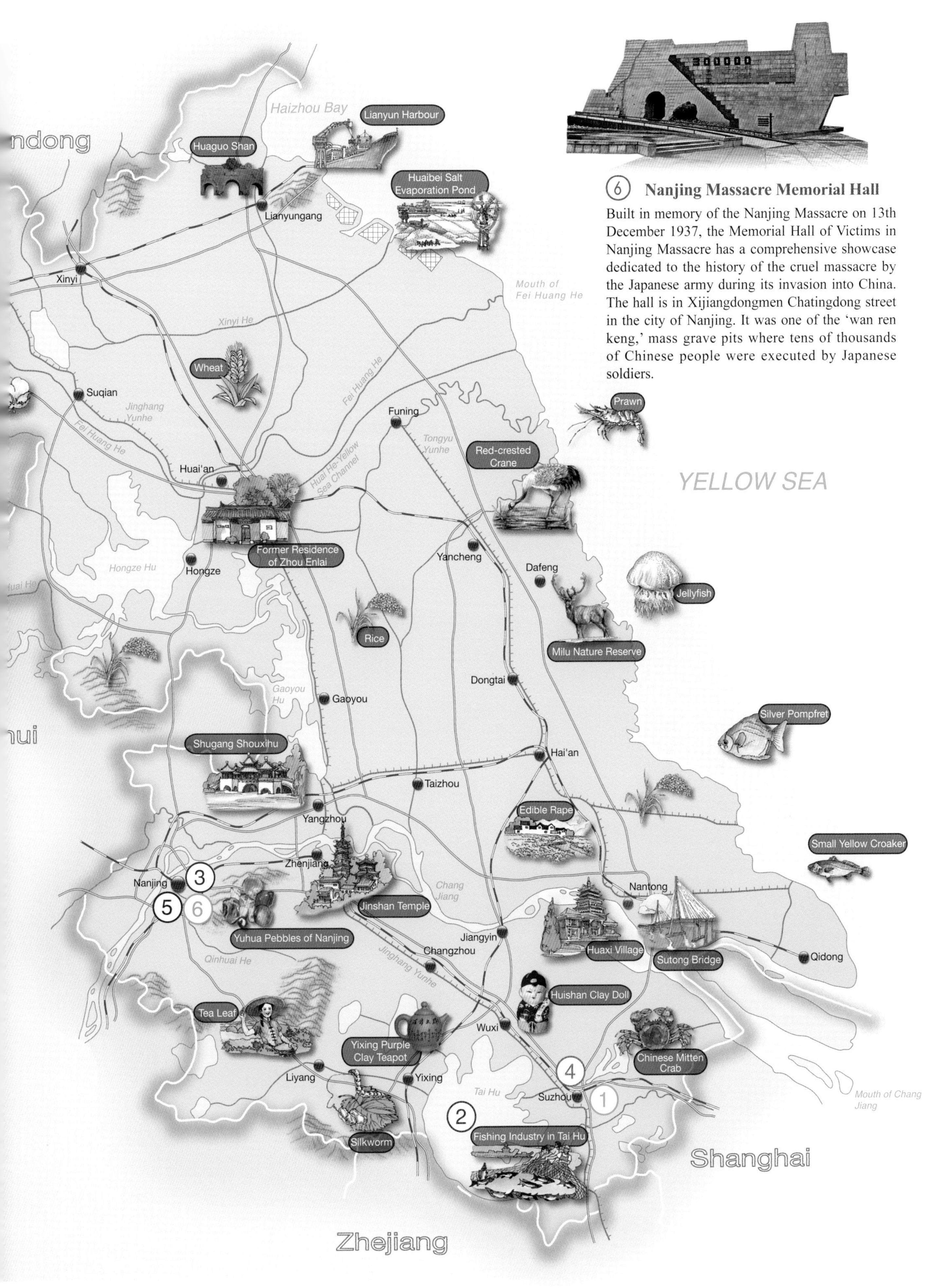

⑥ Nanjing Massacre Memorial Hall

Built in memory of the Nanjing Massacre on 13th December 1937, the Memorial Hall of Victims in Nanjing Massacre has a comprehensive showcase dedicated to the history of the cruel massacre by the Japanese army during its invasion into China. The hall is in Xijiangdongmen Chatingdong street in the city of Nanjing. It was one of the 'wan ren keng,' mass grave pits where tens of thousands of Chinese people were executed by Japanese soldiers.

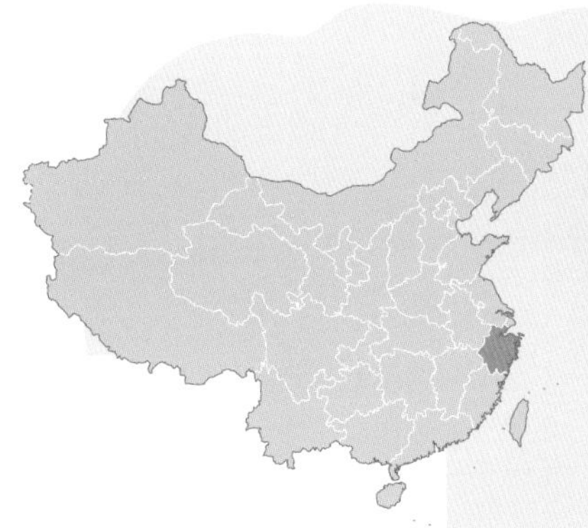

Zhejiang Province

The name Zhejiang, abbreviated as Zhe, originated from Qiantang Jiang, the largest river in the region. The river bends along its way, hence given the name Zhejiang which later became the name of the province.

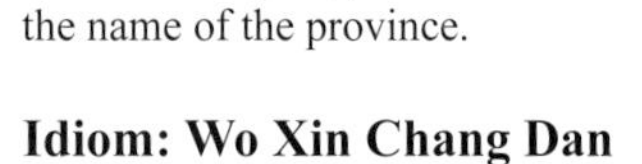

Idiom: Wo Xin Chang Dan

During the Spring and Autumn period, Goujian was the King of the state of Yue, which belonged to Zhejiang with its capital in Huiji (present-day Shaoqing). At that time, Yue was at war against the state of Wu and defeated. To save his State, Goujian endured the humiliation of serving the King of Wu for 3 years. After returning to Yue, he slept on firewood and licked a bitter gall bladder every night to remind himself never to be pleasure-seeking. Under his excellent leadership, Yue grew stronger and finally conquered Wu. Goujian became one of the hegemons of the period.

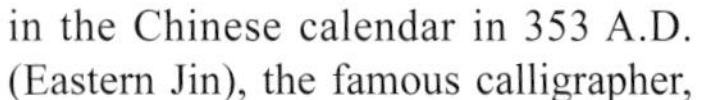

The Origin of *Lantingji Xu*

On the 3rd day of the 3rd month in the Chinese calendar in 353 A.D. (Eastern Jin), the famous calligrapher, Wang Xizhi, and prestigious literati such as Xie An and Sun Zhuo held the elegant social event 'xiuxi' in Lanting Pavilion in Huiji. They made poems and compiled them into *Lantingji*. The preface that Wang Xizhi made for the compilation is crowned as the 'best running script in the world.' The literati gathering by the meandering river at the Lanting Pavilion has been highly acclaimed by later generations.

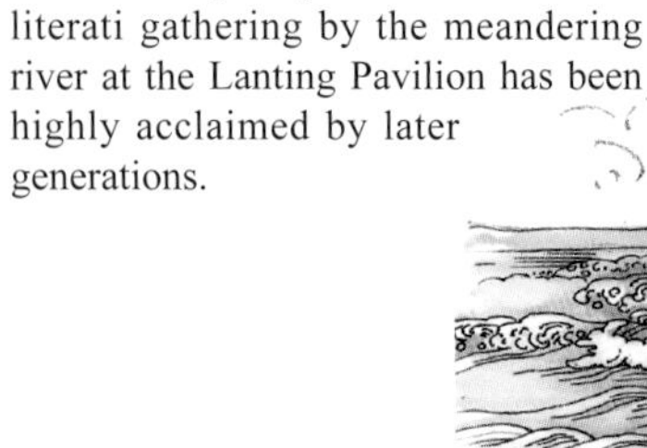

① Xi Hu of Hangzhou

The Song poet Su Shi once praised that the Xihu was comparable to Xi Shi, one of the Four Beauties of Ancient China. Xi Hu, also goes by Xizi Hu, is in the southwest of Hangzhou and an attraction of worldwide fame. The scenery varies in different seasons so that the tourists would be reluctant to part with the beautiful sights. There are also proses, verses, and legends which add to the charm.

② Yueju (Yue Opera)

Originated in the region near Shaoxing, Yueju often features actresses who play male characters in stunning costumes and sing gracefully. Classics include *Dream of the Red Chamber* and *The Butterfly Lovers*.

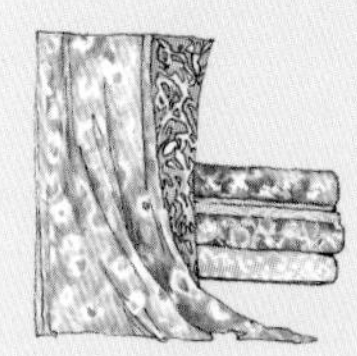

③ Silk of Hangzhou

The silk from Hangzhou is of exceptional quality. It has been transported overseas through the Silk Road as early as in the Han Dynasty.

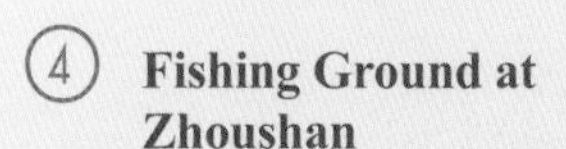

④ Fishing Ground at Zhoushan

The fishing ground at Zhoushan is the largest in China abounding in Dui prawns, large yellow croaker, largehead hairtail, cuttlefish and so on. It is known as the 'Fish Warehouse of the East China Sea.'

⑤ Jinghang Yunhe

Jinghang Yunhe is the longest canal in the world and a famous ancient hydraulic project completed in the Sui Dynasty running from Beijing to Hangzhou.

⑥ Jiangnan Watertown in Shaoxing

Shaoxing is peppered with lakes and latticed by rivers and water alleys. It is a typical example of a Jiangnan watertown.

⑦ High Tide at Qiantang Jiang

The best time to watch the high tide at Qiantang Jiang is around the 18th day of the 8th month in the Chinese calendar at the Tide Watching Park in northern Zhejiang. When the tide surges, it can reach several metres as it rages into the Hangzhou Bay with thundering crush and mighty momentum.

Jiangsu
Shanghai
Tai Hu
Jinghang Yunhe
Huzhou
Huzhou Brush
Silkworm
Jiaxing
Anji Bamboo Sea
Anji
Donghai Bridge
Shengsi
Shengsi Liedao
Yangshan Port
Jiaxing Nan Hu
Hangzhou Bay Bridge
Hangzhou Bay
5
Yanguan
4
Large Yellow Croaker
Tianmu Shan
3
7
Hangzhou
Qiantang Jiang
1
Zhoushan Dao
Cixi
Zhoushan Qundao
Fuyang
6
Shaoxing
Cixi Chinese Treeberry
Zhoushan
Ningbo Port
Ningbo
Beilun Port
Shaoxing Huadiao Rice Wine
Fuchun Jiang
Puyang Jiang
Cao'e Jiang
Siming Shan
Putuo Shan
ea Leaf
Tonglu
Zhuji
Fenghua
Xiangshan Port
Shengzhou
2
Small Yellow Croaker
Former Residence of Xishi
Peach
Jiande
Rice
Dongyang
Huiji Shan
Ninghai
EAST CHINA SEA
Qu Jiang
Jinhua
Hengdian World Studios
Hairtail Fish
Yongkang
Rice
Jinhua Ham
Edible Rape
Linhai
Ling Jiang
Taizhou
Taizhou Bay
Kuocang Shan
Lishui
Huangyan Mandarin Orange
Taizhou Liedao
Cuttlefish
Yandang Shan
Rice
Qingtian
Yuhuan Dao
Yandang Shan
Wenzhou
Wenzhou Bay
Longquan Blue Porcelain
Taishun Lounge Bridge
Prawn
Taishun
Cangnan
Nanjishan Liedao
Fujian

Anhui Province

The name Anhui came from the two main cities in the province, Anqing and Huizhou. Anqing was the land of state of Wan, which became the abbreviation of the province. But some say that the name 'Wan' came from Tiangui Shan in the province, which was called Wangong Shan.

The Place Where Xiang Yu killed himself

Xiang Yu and Liu Bang were heroes from the turmoil of the final years of the Qin Dynasty. In 202 B.C., Xiang Yu, the Hegemon-King of Chu, was defeated by Liu Bang and killed himself at the age of 30 in Wu Jiang, now in Wujiang Town. His story has become the origin of many idioms.

The Battle of Fei Shui

The Battle of Fei Shui is a famous example of strategic wisdom that the weak defeat the powerful. The ancient battlefield was on the shoreside of Fei Shui in central Anhui. In 383 A.D., the ruler of the former Qin Dynasty wanted to unite China. Therefore they sent troops to Jin and fought against the Eastern Jin army in Chenbing near Fei Shui. Finally, the Eastern Jin army of 80,000 soldiers defeated the former Qin army of more than 800,000. The battle has inspired many idioms.

① Huang Shan

In southern Anhui, it is most famous for the bizarre pine trees and rocks, sea of clouds, and springs. There is a saying that 'One would not want to see other mountains after visiting the Five Great Mountains, and would not want to see the Five after visiting Huang Shan.'

② Huangmeixi (Huangmei Opera)

Huangmeixi, is a famous local genre of Chinese opera. The singing style is mild and graceful. Classical opera themes include *The Cowherd and the Weaving Girl.*

③ Wannan Folk Residence

The most typical examples of Wannan folk residence are Xidi Village and Hong Village. Most of the houses have white walls and roofs made of black tiles, sitting beside green mountains and clear water and forming a classical and elegant scenery. With a long history and grand size, the residence is called the 'Museum of Residence of Ming-Qing dynasties.'

④ Chinese Alligator

Mainly living in southern Anhui, Chinese alligator is a rare species in China and now under first-grade state protection.

⑤ Jiuhua Shan

Situated in southern Anhui, Jinhua Shan is one of China's Four Sacred Mountains of Buddhism and honoured as the 'Mountain of the Buddhist Kingdom.'

⑥ Four Treasures of the Study

The four essential tools in a study of ancient literati are: the writing brush, the ink stick, the paper, and the ink stone. The tools are widely varied, but the most famous ones are the Hu writing brush, Hui ink stick, Xuan paper, and Duan ink stone. Hui ink stick and Xuan paper are produced in Anhui.

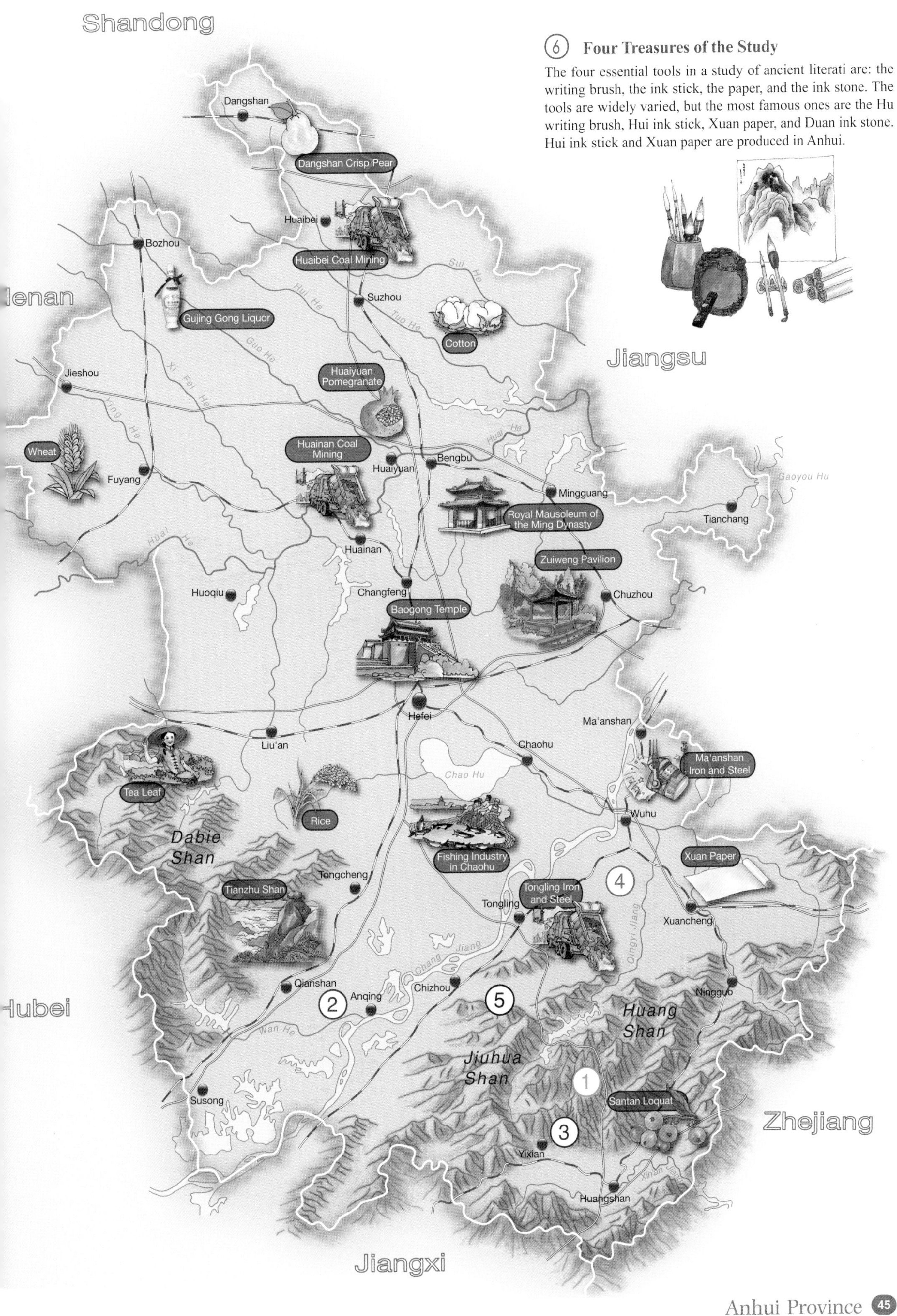

Fujian Province

In 733 A.D. (Tang), the government picked the first word of Fuzhou and Jianzhou and established the Fujian Military Commissioner. This was the first time the name Fujian appeared in history. 'Min' was the collective name of the seven tribes living in the region during pre-Qin period, and later became the abbreviation of Fujian.

Silk Road on the Sea

Silk Road on the Sea, also goes by 'china route on the sea' or 'perfume route on the sea,' formed in Qin and Han dynasties and became popular in Tang and Song dynasties. Its furthest ends reached western Asia and East African coast. It is the oldest route on the sea. The port of Quanzhou in Fujian, also reputed as the 'Largest Port in the Orient,' was the starting point of Silk Road on the Sea in Song and Yuan dynasties.

Opening of the Five Ports

Fujian is one of the first provinces to be opened for trade in contemporary history. In 1842, the Qing government signed the Nanjing Treaty with Great Britain. It designated five ports to be opened for foreign trade with British merchants: Guangzhou, Xiamen, Fuzhou, Ningbo, and Shanghai. Two of them, Fuzhou and Xiamen, are in Fujian province.

① Wuyi Shan

Located in northern Fujian, Wuyi Shan is mainly in Danxia landforms with emerald water and red mountains, curving peaks and meandering streams. Wuyi Shan is a UNESCO World Heritage site and renowned for having the 'unique beauty that tops the scenery of Southeast China.'

② Longan (or Guiyuan)

Longan is a sweet, nutritious fruit. It is also the speciality of Fujian, the top longan producing province in China.

③ Matsu Temple

Matsu Temple in Putian is a monument of Matsu. According to the legend, she had the ability to save people from disasters and has been worshipped as the Sea Goddess.

④ Oolong Tea of Fujian

The tea of Fujian is known internationally and Oolong tea is a particularly famous genre of Fujian tea. Tie Guanyin is the best variety of Oolong tea. It is good for health and known as the 'Tea for Longevity.'

Jiang
Wuyi Shan
Changding
Liancheng
Forest
Shanghang
5
Guangdong

⑤ Tulou of Kejia

Located in Yongding County in southern Fujian, Tulou are traditional houses of Kejia people. They are strong, safe for residence, and unique in shape.

⑥ Gulangyu

Gulangyu, also called the 'Garden on the Sea,' is a beautiful small island in the sea south-western of Xiamen.

Jiangxi Province

Jiangxi means the western part of Jiangnan. In 733 A.D. (Tang), Emperor Xuanzong named the province after the Jiangnan Xidao (Jiangnan West Circuit) built in the region. It is abbreviated as Gan, which is the name of the largest river in Jiangxi.

Chaisang—Political Centre of Jiangdong

Chaisang is the present-day Jiujiang in Jiangxi. In the final years of the Eastern Han Dynasty, Chaisang was the political centre of Jiangdong. During the Three Kingdoms period, Sun Quan came here for state affairs. Before the Battle of Chibi, Zhuge Liang who represented Liu Bei also came to Chaisang to meet Sun Quan to ally with him against Cao Cao. The scene in which 'Zhuge Liang disputed with the southern scholars' is well depicted in the *Romance of the Three Kingdoms*.

Pipa Xing by Bai Juyi

Bai Juyi was a great poet in the Tang Dynasty. He wrote *Pipa Xing* when being banished to Jiujiang Prefecture as a low-ranking executive officer. Jiujiang Prefecture is located in present-day Jiujiang in Jiangxi. Other places such as Xunyang Jiang and Jiangzhou in the poem are also in Jiujiang.

① Teng Wang Pavilion

Standing beside Gan Jiang in Nanchang, Teng Wang Pavilion is one of the Three Great Towers in Jiangnan together with Yueyang Tower in Hunan and Huanghe Tower in Hubei. It has become a legend because of the classic prose *Tengwang Ge Xu* written by Wang Bo more than 1,300 years ago.

② Jingdezhen Porcelain

The porcelain manufacturing industry in Jingdezhen has over 2,000 years of history. Known for its exceptional qualities, the porcelain represents the highest level of porcelain craftsmanship in China. That is why Jingdezhen is reputed as the 'Porcelain City.'

③ Wuyuan — China's Most Beautiful Village

Situated in north-eastern Jiangxi, Wuyuan is well-known for the elegant field scenery depicted in pastoral songs and the well-conserved ancient residence. The best time to visit 'China's Most Beautiful Village' is when the golden rape flowers are in full blossom in spring.

④ Poyang Hu — China's Largest Freshwater Lake

Poyang Hu is a vast expanse of mist-covered water with abundant water products. It is the largest fresh water lake in China and a paradise for migratory birds as millions of them come to pass the winter.

⑤ Lu Shan

Situated in northern Jiangxi, Lu Shan lies in the south of Chang Jiang and the west of Poyang Hu. Numerous literati across dynasties praised in prose and verse about the elegant peaks, sea of clouds, rushing waterfalls, and flowing springs there. The most famous example is Li Bai's poem *Wang Lushan Pubu*.

Shandong Province

Located in the east of Taihang Shan, Shandong is given the name meaning the 'east of mountain.' It is abbreviated as Lu or Qilu. After Emperor Wu of Zhou defeated the Shang Dynasty, he gave the land of Qi to Jiang Ziya and the land of Lu to his brother Zhou Gong. Both Qi and Lu are in present-day Shandong and thus become the abbreviations of the province.

Ancient Battlefield of the Battle of Changshao

The Battle of Changshao was in present-day Laiwu and a typical example of strategic wisdom in Chinese history. In 684 B.C., Duke Huan of Qi sent troops to invade the State of Lu. When the two armies met in Changshao, the Qi army beat the drum of war as a signal of attack, but the Lu army, led by Duke Zhuang of Lu and Cao Gui, did not react at all. After the Qi army beat the drum for three times, the Lu army suddenly drummed and attacked the dispirited Qi army. Finally the Lu army won the battle. The battle is also the origin of the idiom Yi Gu Zuo Qi.

Weihai and the Beiyang Fleet

When the First Sino-Japanese War broke out in 1894, the Beiyang Fleet was defeated in the Battle of Yellow Sea and withdrew to Liugong Island in Weihai to reorganize. In January of 1895, the Japanese army attacked Weihai from on land, and sent fleet to block the sea of Weihai. Attacked from the front and rear, the Beiyang Fleet was forced to surrender and totally defeated.

① Confucian Temple in Qufu

Qufu is the birthplace of Confucius. The grand and splendid Confucian Temple is where Chinese emperors worship him.

② Kites of Weifang

Having a long history and exceptional quality, kites made in Weifang are internationally famed handicraft. Every April the International Kite Festival is held in Weifang. There is also the Weifang Kite Museum that displays all sorts of kites from around the world.

Hebei

Dezhou

Dezhou Watermelon

Maize

Linqing

Yucheng

Jinghang Yunhe

Liaocheng

Jinan

Feicheng Peach

Feicheng

Dongping Hu

Henan

Shanxian

Lotus

Anhui

③ Seaside City of Qingdao

Qingdao combines the scenery of mountains, sea, and city as concluded in the Chinese saying 'red tile roofs among verdurous trees, the lagoon-blue sky above the sapphire sea.' It is a world-famous seaside resort for escaping heat.

④ Tai Shan

Tai Shan is reputed 'the Most Prominent among China's Five Great Mountains.' Commanding a spectacular view, Tai Shan is also home to many Buddhist relics and enlisted as a UNESCO World Heritage site.

⑤ Confucius

Confucius is a great educator, philosopher, and the founder of Confucianism. His thoughts and speeches are recorded in the *Analects* which has had a profound influence on Chinese civilisation.

① Shaolin Temple in Song Shan

Song Shan is one of China's Five Great Mountains. Shaolin Temple built in Song Shan is the most legendary ancient temple of in China. Its martial arts have fascinated people home and abroad.

Henan Province

With most of its regions in the south of Huang He, it is given the name Henan which literally means 'south of river.' It is abbreviated as 'Yu' as it shares similar administrative division with Yuzhou, one of the legendary 'Nine Provinces' in the ancient document *Yugong*.

The Battle of Muye

The Battle of Muyue was a major battle that the Zhou Dynasty overthrew the Shang Dynasty. Emperor Zhou of Shang was a ruthless and lustful tyrant in the later years of the Shang dynasty. The country was very weak under his rule. In 1046 B.C. (Shang), Emperor Wu of Zhou gathered the support of other tribes and multi-state states. They formed an army and march to the Shang capital, Chaoge (present-day Qixian). When they reached Muye, which was very close to Chaoge, the major force of the Shang army was fighting in the southeast far away from the capital. Emperor Zhou hurriedly gathered about 700,000 people comprising slaves, prisoners of wars, and guards of the capital to fight in Muye. As the battle started, the Shang army had no will to fight and quickly collapsed. Knowing that he was bound to lose, Emperor Zhou returned to Chaoge and burned himself to death on Lutai. The Shang Dynasty thus ended.

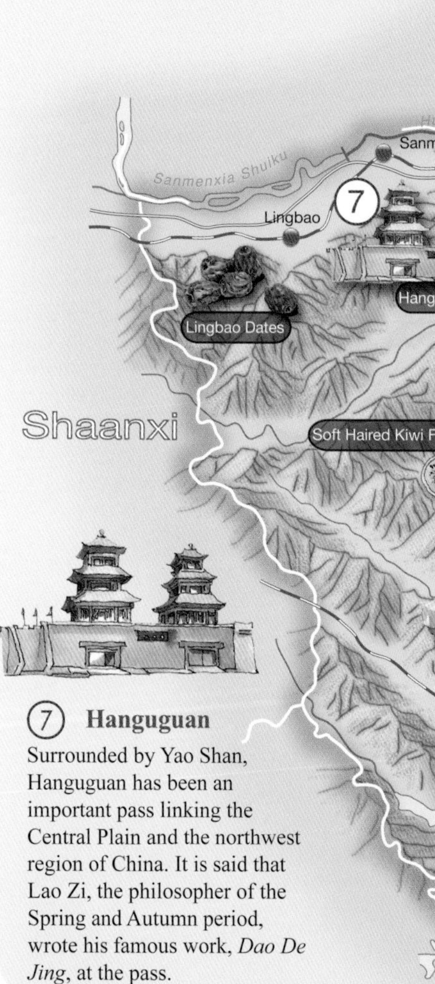

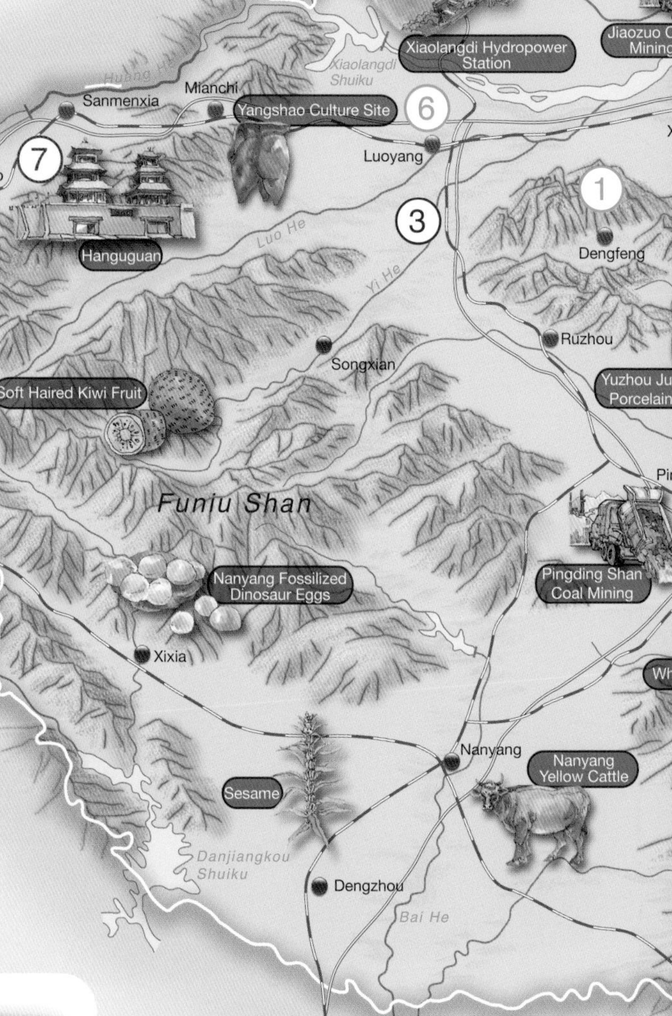

⑦ Hanguguan

Surrounded by Yao Shan, Hanguguan has been an important pass linking the Central Plain and the northwest region of China. It is said that Lao Zi, the philosopher of the Spring and Autumn period, wrote his famous work, *Dao De Jing*, at the pass.

② Riverside Scenery at the Qingming Festival

The wide hand scroll painting of *Riverside Scenery at the Qingming Festival* is a national treasure drawn by Zhang Zeduan, a painter of the Northern Song Dynasty. It depicts the scene of prosperity in Kaifeng of Henan during Qingming Festival of the Northern Song Dynasty more than 900 years ago.

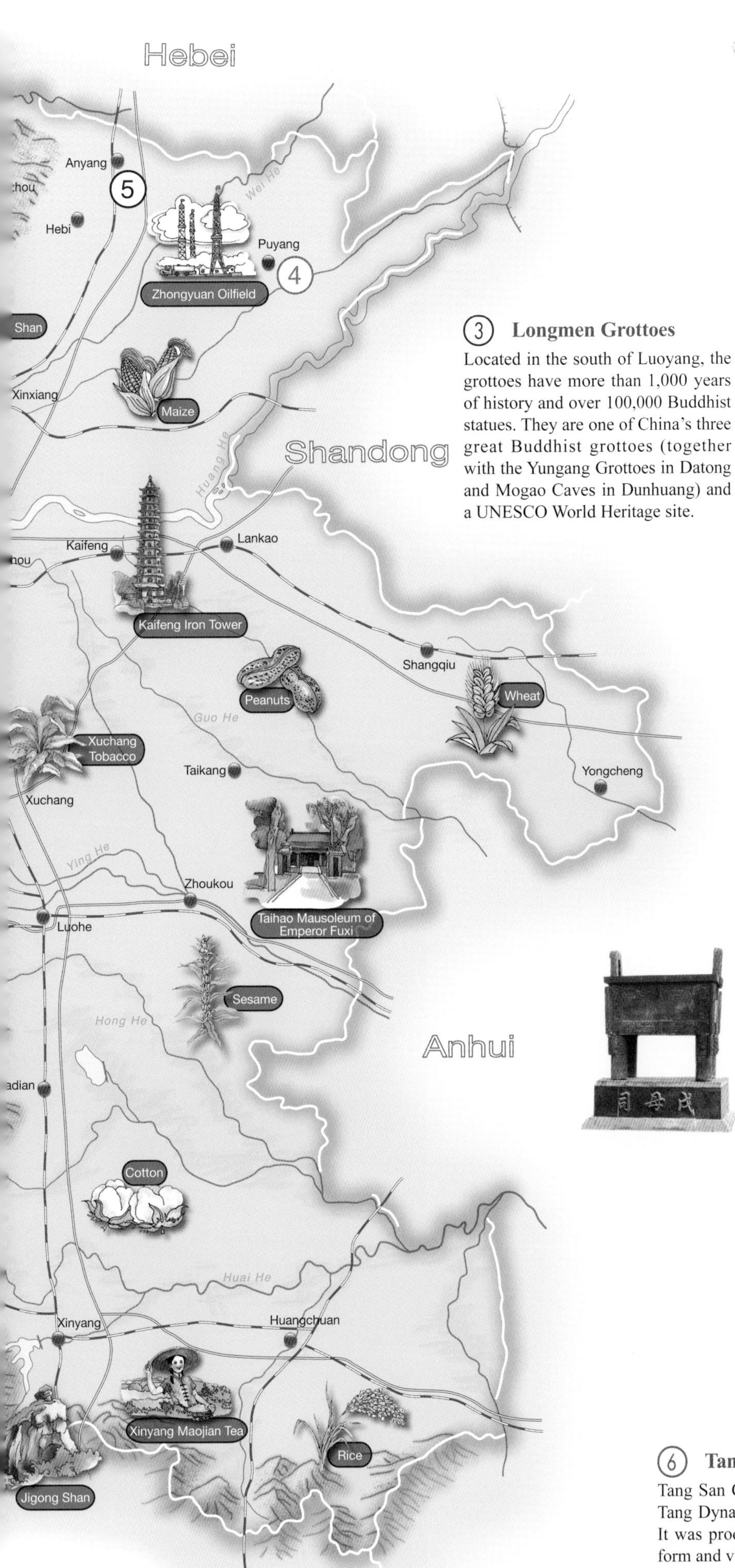

③ Longmen Grottoes

Located in the south of Luoyang, the grottoes have more than 1,000 years of history and over 100,000 Buddhist statues. They are one of China's three great Buddhist grottoes (together with the Yungang Grottoes in Datong and Mogao Caves in Dunhuang) and a UNESCO World Heritage site.

④ First Dragon of China

In 1987, a dragon-tiger totem made up of oyster shells was found in the remains of Yangshao Culture in the northeast of the province. The dragon is 1.78 m long and 0.67 m tall. Archaeologists confirmed that it was from 6,000 years ago and the 'First Dragon of China.'

⑤ Yinxu Site in Anyang

Yinxu is the historical site of the capital of late Shang Dynasty with more than 3,300 years of history. It was famous for the abundant discoveries of the oracle bone script and bronzeware. Among them, the Simu Wu (or Houmu Wu) Ding vessel is a national treasure.

⑥ Tang San Cai

Tang San Cai is a tri-coloured glazed pottery popular in the Tang Dynasty. Its main colours are cream, amber, and green. It was produced mainly in Luoyang and Xi'an with lifelike form and vivid colours.

Hubei Province

The name Hubei, meaning 'north of lake,' is given as the province lies north of Dongting Hu, Wuchang has become its administrative centre since the Qing Dynasty. From the Sui Dynasty onwards, Wuchang has been the location of the governing body of Ezhou (also called Jiangxia). Therefore, 'E' has become the abbreviation of Hubei.

The Battle of Chibi

The ancient battlefield of Chibi is in present-day Chibi in south-western Hubei. In 208 A.D., Cao Cao led an army of 200,000 southward to attack Jingzhou against the union armies of Sun Quan and Liu Bei on opposite sides of the river. Initially, the Sun-Liu armies were in an inferior position. They knew Cao's army was not good at fighting on water and not used to the weather, so they attacked Cao's army with fire and won the battle. The battle set up the tripartite confrontation in China.

Wuhan Uprising in the Xinhai Revolution

On 10 October 1911 (the year of Xinhai in the Chinese calendar), revolutionists started the 'First Uprising in Wuchang' which sparked off the Xinhai Revolution. In the following two months, 15 provinces including Hunan and Guangdong announced their separation from the Qing government. On 1st January 1912, the Republic of China was set up in Nanjing and Sun Yat-sen was named the provisional president. On 12th February, the resignation of Puyi, the last emperor of Qing, marked the end of the Qing Dynasty. More than 2,000 years of feudalism was ended in China and the Republic of China was set up.

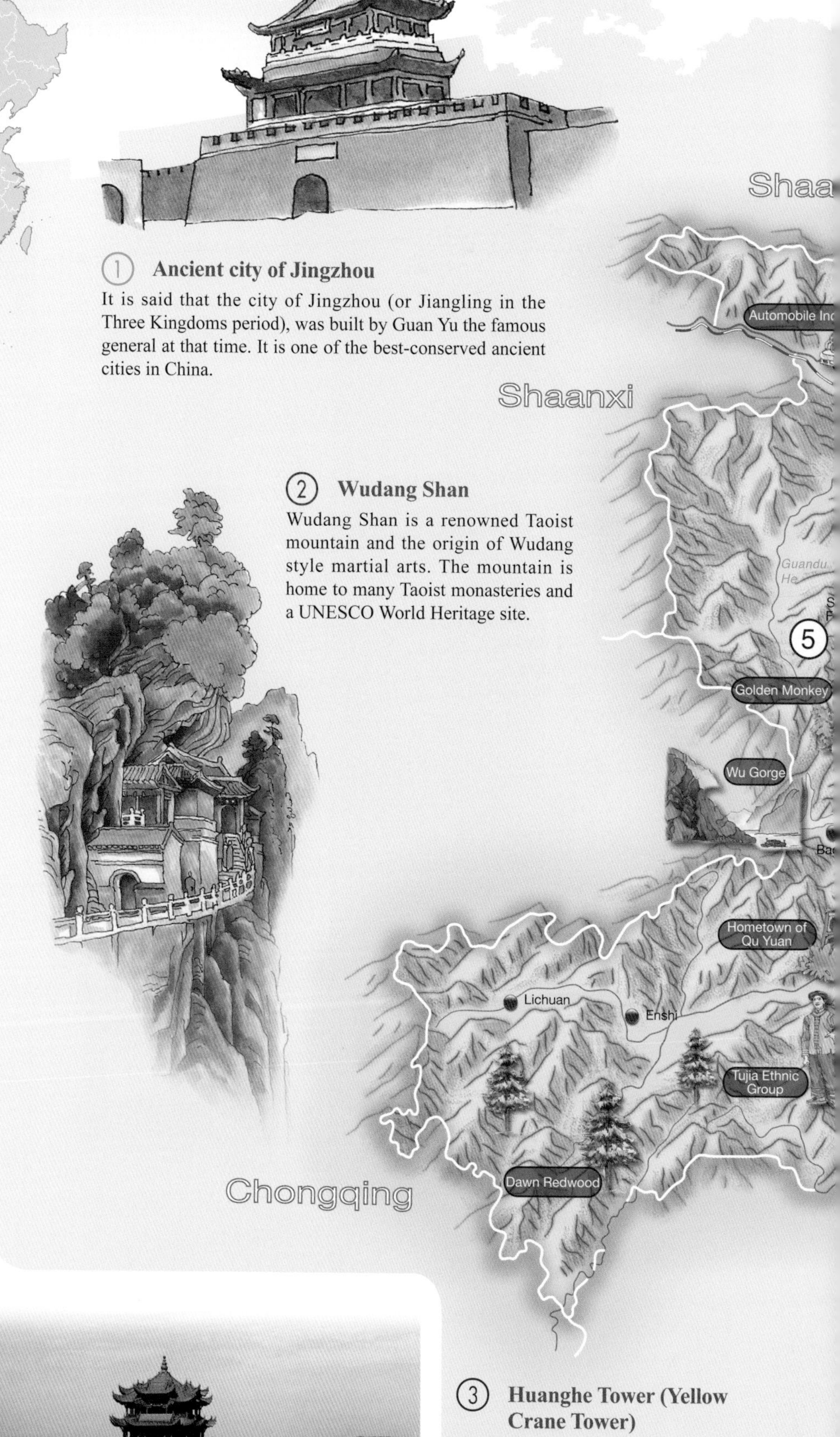

① Ancient city of Jingzhou

It is said that the city of Jingzhou (or Jiangling in the Three Kingdoms period), was built by Guan Yu the famous general at that time. It is one of the best-conserved ancient cities in China.

② Wudang Shan

Wudang Shan is a renowned Taoist mountain and the origin of Wudang style martial arts. The mountain is home to many Taoist monasteries and a UNESCO World Heritage site.

③ Huanghe Tower (Yellow Crane Tower)

Huanghe Tower was built in the Three Kingdoms period. Situated on She Shan in Wuhan by Chang Jiang, it is one of the Three Great Towers of Jiangnan together with Yueyang Tower in Hunan and Tengwang Pavilion in Jiangxi. For hundreds of years, numerous literati have visited the tower and left proses and verses there.

⑦ Bianzhong (a set of sequenced bells) of Marquis Yi's Tomb

Located in Suizhou, the tomb belonged to the ruler of the State of Zeng in early Warring States period. The whole set of Bianzhong covers five and a half octaves and all the twelve semitones. With exquisite casting skills and good musical quality, the Bianzhong rewrote the music history of the world and earned the reputation of 'an extremely rare treasure.'

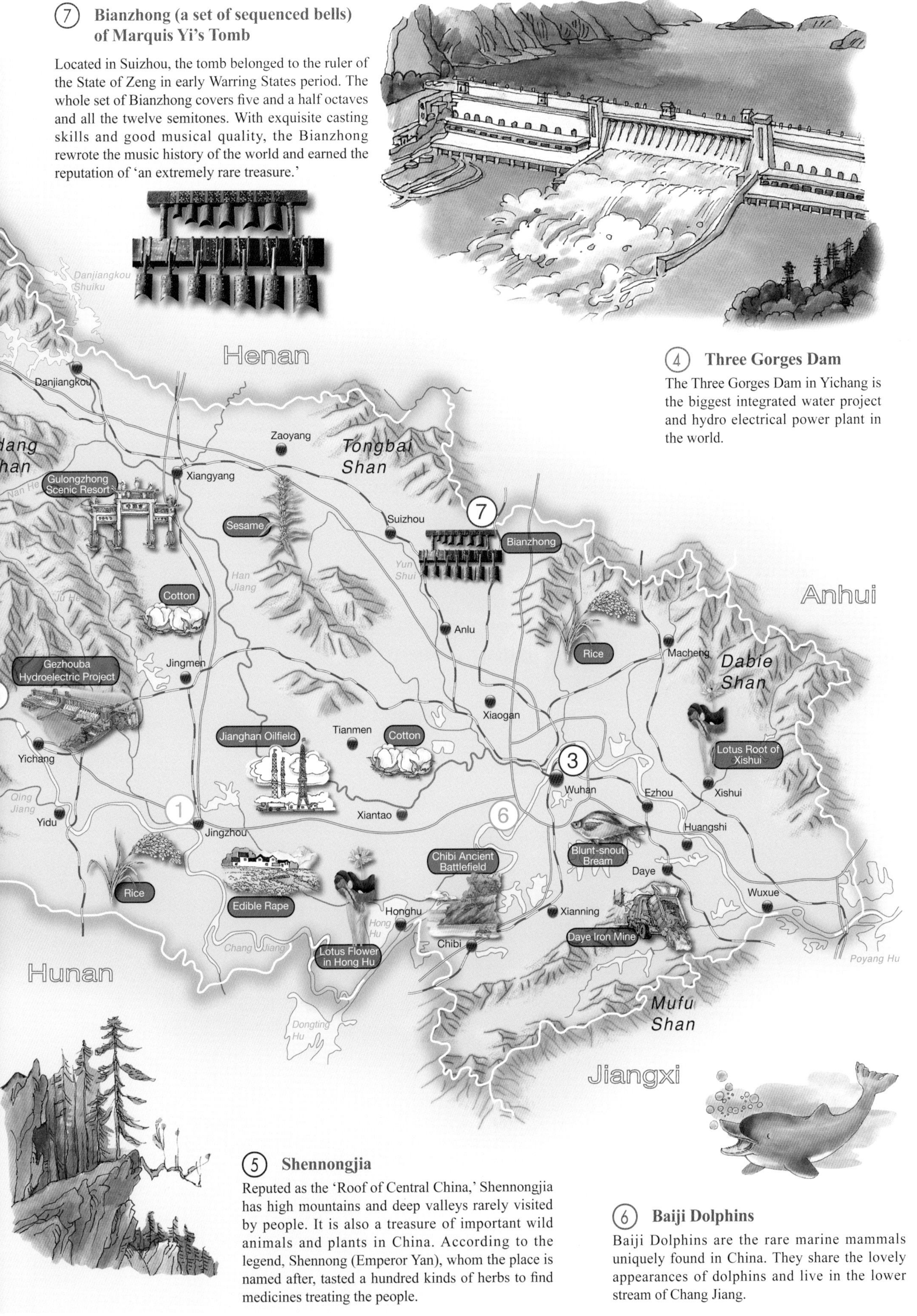

④ Three Gorges Dam

The Three Gorges Dam in Yichang is the biggest integrated water project and hydro electrical power plant in the world.

⑤ Shennongjia

Reputed as the 'Roof of Central China,' Shennongjia has high mountains and deep valleys rarely visited by people. It is also a treasure of important wild animals and plants in China. According to the legend, Shennong (Emperor Yan), whom the place is named after, tasted a hundred kinds of herbs to find medicines treating the people.

⑥ Baiji Dolphins

Baiji Dolphins are the rare marine mammals uniquely found in China. They share the lovely appearances of dolphins and live in the lower stream of Chang Jiang.

Hunan Province

Hunan, literally meaning 'south of lake,' is given the name as most of the region lies to the south of Dongting Hu. Its abbreviation 'Xiang' is named after Xiang Jiang that flows through the province.

Dragon Boat Festival and Hunan

According to the legend, Dragon Boat Festival is in memory of Qu Yuan, a great poet from the state of Chu in the final years of the Warring States period. Despite being loyal to the state of Chu and the people, he was shunned by vicious officials and banished to Dongtian Hu by the Emperor of Chu. In 278 B.C., the state of Qin occupied Ying, the Chu capital. Qu Yuan felt hopeless about Chu and jumped into the Miluo Jiang on the 5th day of the 5th month in the Chinese calendar. It has become a custom to hold dragon-boat contests and wrap rice dumplings to lament his death.

Zeng Guofan and the Xiang Army

Xiang Army is a local army set up in Hunan in late Qing Dynasty. As the Qing army could not resist the Taiping Rebellion, the government had to seek help from local forces. At this time, Zeng Guofan set up and expanded the Xiang Army. Later, leaders of the Xiang Army and their chancellors played major roles in making political and military decisions for China. They were significant to contemporary history of China.

① Wulingyuan

Located in Zhangjiajie in western Hunan, Wulingyuan is reputed as a gift from nature. It is a perfect blend of countryside scenery and lush woodlands, comprising jagged stone pillars, deep valleys, limestone caves with winding tunnels, clear streams, and luxuriant, dense forests.

② Mao Zedong's Former Residence

Xiangtan is the birthplace of Mao Zedong (1893—1976), founder of the People's Republic of China. He also grew up as a teenager there.

Chongq

③ Buddhist Gauze Silk Robe from Mawangdui Han Tomb

The Mawangdui Han Tomb is in the countryside in eastern Changsha. It belonged to the family of Li Cang, the Marquis of Dai and the minister of the state of Changsha in early Western Han Dynasty. There are more than 3,000 precious items, among which a Buddhist gauze silk robe is exceptionally light and thin—the robe is 1.28 m long but weighs only 49 g.

④ Dongting Hu — Yueyang Tower

Dongting Hu is the second largest freshwater lake in China. Yueyang Tower, standing beside the lake, is one of the Three Great Towers in Jiangnan together with Huanghe Tower in Wuhan and Tengwang Pavilion in Nanchang. It has become well-known after Fan Zhongyan wrote the famous prose *Yueyanglou Ji*.

Guizh

⑤ Heng Shan

Situated in central Hunan, Heng Shan is the southernmost and the 'Most Elegant among China's Five Great Mountains.' It consists of giant, old trees, and 72 peaks.

⑥ Hanging Dwellings in Western Hunan

Hanging dwellings are common residence in Western Hunan. The houses are designed to be dry, well-ventilated, and land-saving.

Hubei
Shimen
Lixian
Li Shui
Edible Rape
Dongting Hu Fishing Industry
Yueyang
Chang Jiang
Dongting Hu
Tea Leaf
1
Zhangjiajie
Changde
Yongshun
Tujia Ethnic Group
Wuling Shan
Yuan Jiang
Taohuayuan Resort
Hunan Lotus Seed
Miluo Jiang
Yiyang
Southern Great Wall
Navel Orange
Zi Shui
Xiangxiu Embroidery
Liuyang Firework
Jishou
Yuelu Academy
Changsha
Liuyang
Xuefeng Shan
Fenghuang
6
Xupu
2
Shaoshan
3
Fenghuang Ancient City
Lengshuijiang Antimony Mining
Loudi
Lengshuijiang
Xiangtan
Zhuzhou
Liling
Huaihua
Yuan Jiang
Jiangxi
Shaoyang
5
Xiang Jiang
Dongkou
Zi Shui
Youxian
Hengyang
Mi Shui
Luoxiao Shan
Rice
Emperor Yan Mausoleum
Wuxi Beilin Scenic Resort
Qiyang
Nanshan Pasture
Jingzhou
Yanling
Yongzhou
Leiyang
Miao Ethnic Group
Lei Shui
Shuikoushan Lead and Zinc Mining
Xiao Shui
Guangxi Zhuangzu Zizhiqu
Chongling Shui
Chenzhou
Nan Ling
Dongjiang Shuiku
Daoxian
Yao Ethnic Group
Guangdong

Guangdong Province

In the Song Dynasty, Guangnan Donglu ('lu' was a local administrative unit at that time) was set up with Guangzhou as its capital. Guangdong is the abbreviation of Guangnan Donglu. The province is abbreviated as Yue which initially meant Baiyue region in southern China.

Su Shi Wishing to Stay in Southern China

Luofu Shan in Guangdong, also named Dongqiao Shan, is one of China's Ten Famous Daoist Mountains after Ge Hong from the Jin Dynasty practised Daoism and wrote the Daoist classic *Baopuzi* there. Luofu Shan is also called the Most Famous Mountain in Lingnan region. In the Northern Song Dynasty, Su Shi (also called Su Dongpo) was a literary man banished to Lingnan. He arrived there and wrote the poem *Huizhou Yi Jue*, expressing his wish to stay in Lingnan and his love for the local fruit, lychee. The poem helped spread the fame of Lingnan and Luofu Shan.

① Xinghu Scenic Area

Xinghu Scenic Area is a national-level major scenic site comprising two attractions: Qixing Yan and Dinghu Shan. Qixing Yan, also goes by Xing Hu, is a group of seven rocks shaped like the Plough. Dinghu Shan is an important natural reserve area with a lush forest containing a wide range of plant species.

② Destroying Opium in Humen

In June of 1839, the national hero Lin Zexu ordered to confiscate 2.37 million catties of opium from British and American opium traders and destroy the opium in public on the seaside of Humen in Dongguan. This is the shocking incident of 'Humen Xiaoyan.' The site has been particularly built into the Opium War Museum.

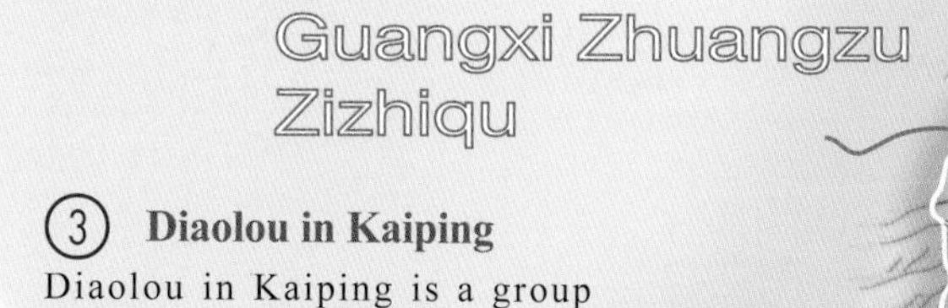

③ Diaolou in Kaiping

Diaolou in Kaiping is a group of local dwellings with unique appearance and included in the *UNESCO World Heritage List*.

④ Shenzhen Special Economic Zone

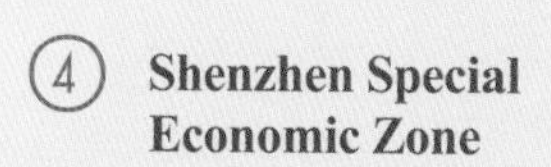

Established in 1980, Shenzhen is regarded as the window of China's reform and open-door policy. Shenzhen has transformed from a tiny, impoverished fishing village into a modern coastal city.

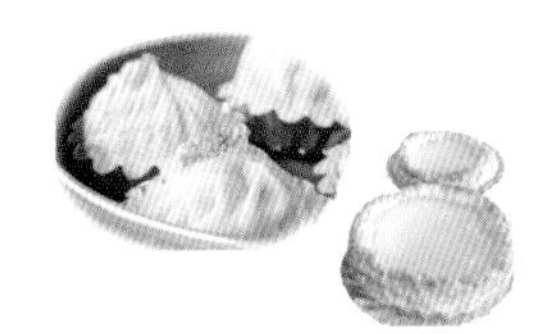

⑤ 'Morning Tea' in Guangzhou

Having 'morning tea' is a typical Guangdong custom and social activity in which people enjoy a wide array of tasty dim sums.

⑥ Five-Goat Statue

The Five-Goat Statue on Yuexiu Shan is a symbol of Guangzhou. It is said that in ancient times there were five immortals with stalks of rice riding on five goats of different colours descended from the sky to Guangzhou. Since then Guangzhou became affluent that no famines have ever happened. After they left, the five goats turned to stone guarding the city. That is why Guangzhou is nicknamed the Goat City.

Dongsha Qundao

⑦ Whampoa Military Academy

In 1924, Dr. Sun Yat-sen established the National Guangdong University (present-day Zhongshan University) and Whampoa Military Academy. Whampoa Military Academy (or the Chinese Nationalist Party's Army Officer Academy), originally set up in Cheung Chau Island in Guangzhou, nurtured many well-known generals and played an important role in Chinese history.

Guangxi Zhuangzu Zizhiqu

In 214 B.C., the Qin Dynasty set up the prefectures of Guilin, Nanhai, and Xiang in Lingnan region. Guilin and Xiang cover most parts of present-day Guangxi. In the Song Dynasty, 'Guangnan Xilu,' later abbreviated as 'Guangxi Lu' were built here. In the early days of Minguo, the capital of Guangxi was Guilin, so 'Gui' has become the abbreviation of the region.

Victory at Zhennanguan

Zhennanguan (present-day Friendship Pass) was at the Sino-Vietnamese border in Pingxiang of south-eastern Guangxi. The battle was a rare victory of the late Qing government against foreign forces. In March of 1885, the French army invaded Zhennanguan. The Qing army was led by a veteran general, Feng Zicai. They fought with full devotion and resolution. Finally, they defeated the French and chased after them into Vietnam. When the defeat was heard in Paris, the majority of the Parliament banned the proposal of additional military expense and the Prime Minister Jules Ferry resigned out of blame. The victorious Qing government wanted peace, and signed the Sino-French Treaty with France on 9th June 1885.

① Fengyu Bridge in Chengyang

Located in Sanjiang Dong Autonomous County, it is the largest and best preserved Fengyu Bridge in China. Combining the bridge, covered corridor, and pavilion structure, it is a rare artistic remains of wooden and stone architecture of the Dong ethnic group.

② Ling Qu

Located in Xing'an County, Ling Qu is a canal that links up the waters of Xiang Jiang and Li Jiang. Its construction started in the Qin Dynasty over 2,200 years ago and it is one the oldest canals in the world.

③ Song Festival on the 3rd Day of 3rd Month

On the annual song festival, the Zhuang people sing mountain songs on which young men and women gather on the field and sing impromptu duets, and the beautiful notes echo through the Zhuang villages.

④ Guilin Landscape

Guilin has the world's most typical karst landscape and is a well-known tourist destination. The beautiful scenery makes it a household name and is best depicted in Han Yu's poem *Song Guilin Yandaifu*.

⑤ **Zhuang Jin**

Zhuang Jin is an elaborate, colourful, traditional handicraft made by Zhuang women. It was presented as a tribute to the emperors in ancient times.

Hainan Province

Hainan Province includes Hainan Dao, South China Sea Islands and their waters. The name comes from Hainan Dao, the administrative centre of the region. In ancient times, it was called Qiong Ya or Qiong Zhou. Thus the region is abbreviated as Qiong.

Extremes of Hainan

Hainan has the largest difference of latitude, the smallest land area and the largest water area among all provinces in China. The total area of the waters is about 2 million km^2 and that of land, 35,000 km^2. The total area of Hainan Dao, also the second-largest island in China, is 33,900 km^2.

Hai Qing Tian

As Hainan is in the farthest end of China, it has long been a prison for exiles. In the Tang Dynasty, a minister Li Deyu was banished to Hainan. after losing his dominance in the 'Niu-Li Factional Struggle' and in the Ming Dynasty, there was an honest and upright official, Hai Rui. His fame was as widespread as Bao Zheng (also called Bao Qing Tian) in the Song Dynasty, and was called 'Hai Qing Tian.' Hai Rui was popular for being upright, determined, righteous, and did not submit to the rich and powerful. He was so popular that stories about him were widespread in folklore. When he died being the censor-in-chief of Nanjing, the locals carried out strikes to mourn him. While his coffin was being transported back to his hometown by boat, the riversides were filled with streams of mourners lamenting his death.

① Coconut Trees

Hainan Dao is the world of coconut forests. In particular, the Dongjiao coconut forest in the countryside of eastern Wenchang is known as the 'Home of Coconut Trees.'

② Rubber Plantation Industry

With a climate that facilitates the plantation of tropical economic plants such as rubber, Hainan becomes the production base of natural rubber plantation.

③ Mangrove Forest in Dongzhai Harbour

Situated in northern Hainan, The mangrove forest in Dongzhai Harbour is China's largest and best conserved mangrove forest and known as the 'Forest on the Sea.'

④ Seaside City of Sanya

Sanya is in the southernmost of Hainan Dao and reputed as the 'edge of the sea and rim of the sky.' It is the most beautiful tropical seaside destination in China. With warm sunlight, soft sand beaches of, clear seawater, fresh salty air, and swaying coconut trees, it is as charming as a paradise.

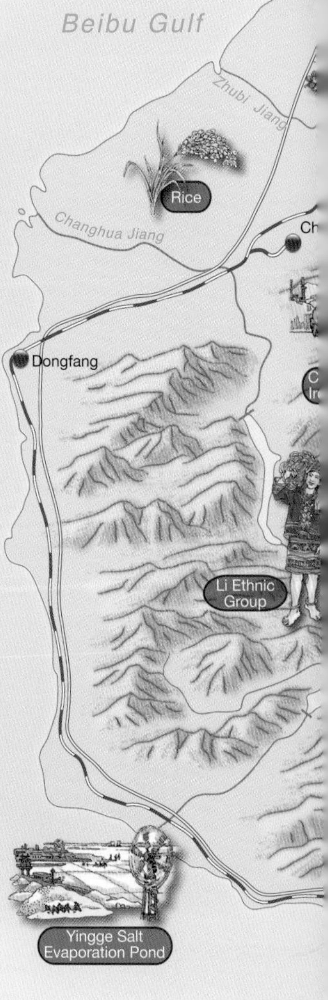

⑤ Li Ethnic Group

Li people are one of China's 56 ethnic groups and in Wuzhishan region in the mid-southern coast of Hainan Dao. They are fond of areca nuts and the bamboo pole dance.

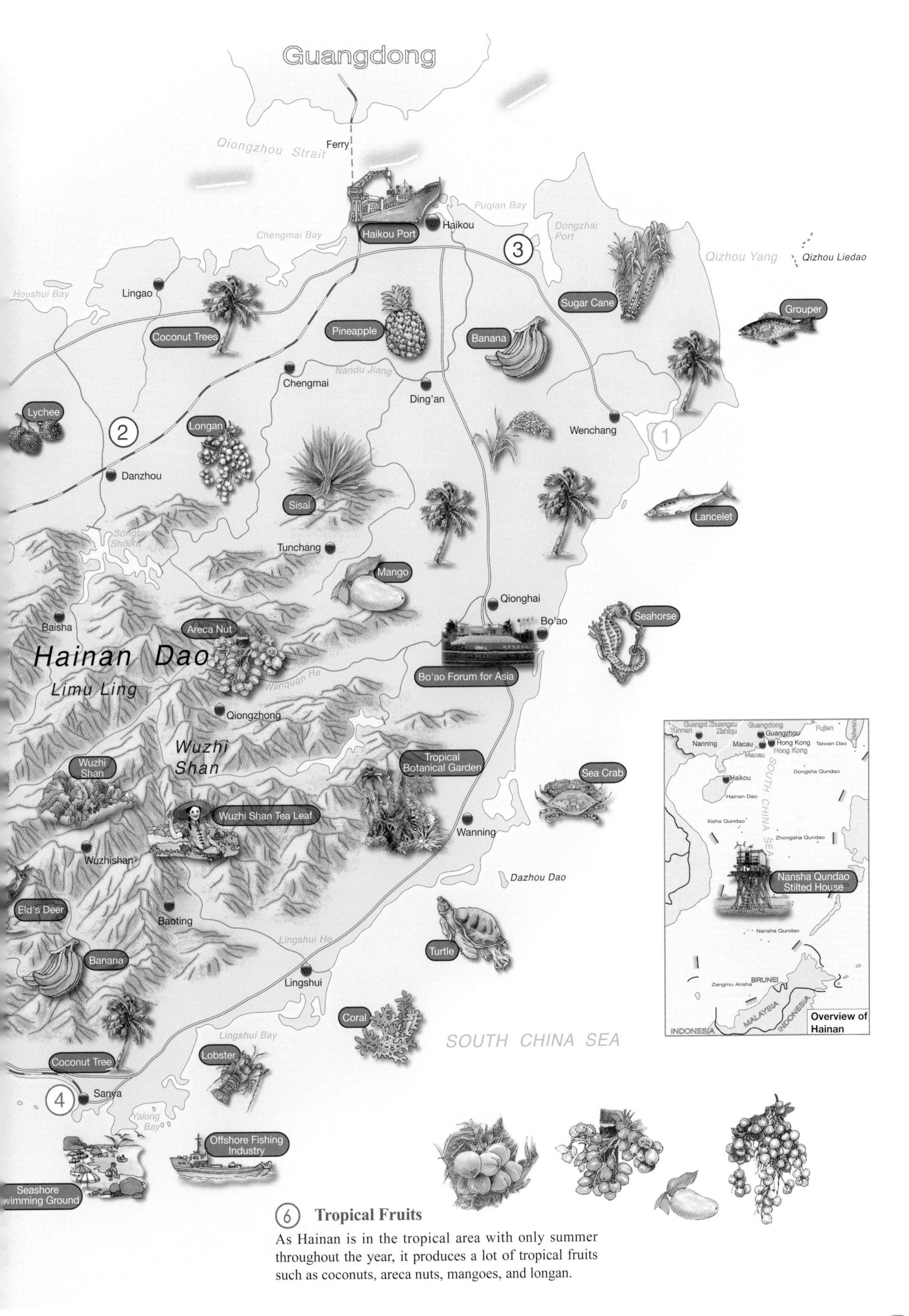

⑥ Tropical Fruits

As Hainan is in the tropical area with only summer throughout the year, it produces a lot of tropical fruits such as coconuts, areca nuts, mangoes, and longan.

Chongqing

In the Sui Dynasty, Yuzhou (present-day Chongqing) was set up beside Jialing Jiang (called Yushui in ancient time). That is why Yu has become the abbreviation of Chongqing. In 1102, Emperor Huizong of Song renamed Yuzhou as Gongzhou. In 1189, Guangzong of Song was at first titled Gongwang and then ascended the throne. To celebrate the good news, he upgraded Gongzhou as Chongqing Fu, meaning 'double celebration.' That is the origin of the name 'Chongqing.'

The Battle of Diaoyu Cheng

Strategically located in Hechuan of Chongqing, Diaoyu Cheng is easy for defence but difficult to invade. In the final years of Southern Song Dynasty, a major battle called the Battle of Diaoyu Cheng broke out there. At that time, Meng Ge, the Khan of Mongolia decided to send strong armies to invade the Song Dynasty. He wanted to put major forces in capturing Sichuan then go east along the river until they reached Lin'an, the capital of Southern Song Dynasty. In 1259 A.D. (Song), the Mongols fought against the Song army in Diaoyu Cheng. But the defence of the Song soldiers and locals was so strong that the Mongols were at their wits' ends. In July of 1259, Meng Ge was killed in a cannon attack so the Mongols had to retreat and stop the invasion. The Battle of Diaoyu Cheng led to another 20 years of rule of the Southern Song Dynasty. It halted the expansion of ancient Mongolia Empire and even changed the setting of the Eurasian battlefield.

(1) Dazu Rock Carvings

Dazu holds tens of thousands of well-preserved Buddhist statues a thousand years ago. The rock carvings were included in the *UNESCO World Heritage List* and considered as the symbol of art of rock carving in Tang and Song dynasties.

(2) Hotpot in Chongqing

Hotpot in Chongqing is popular for its uniquely numb and exciting flavour. It has become the most famous local speciality in Chongqing.

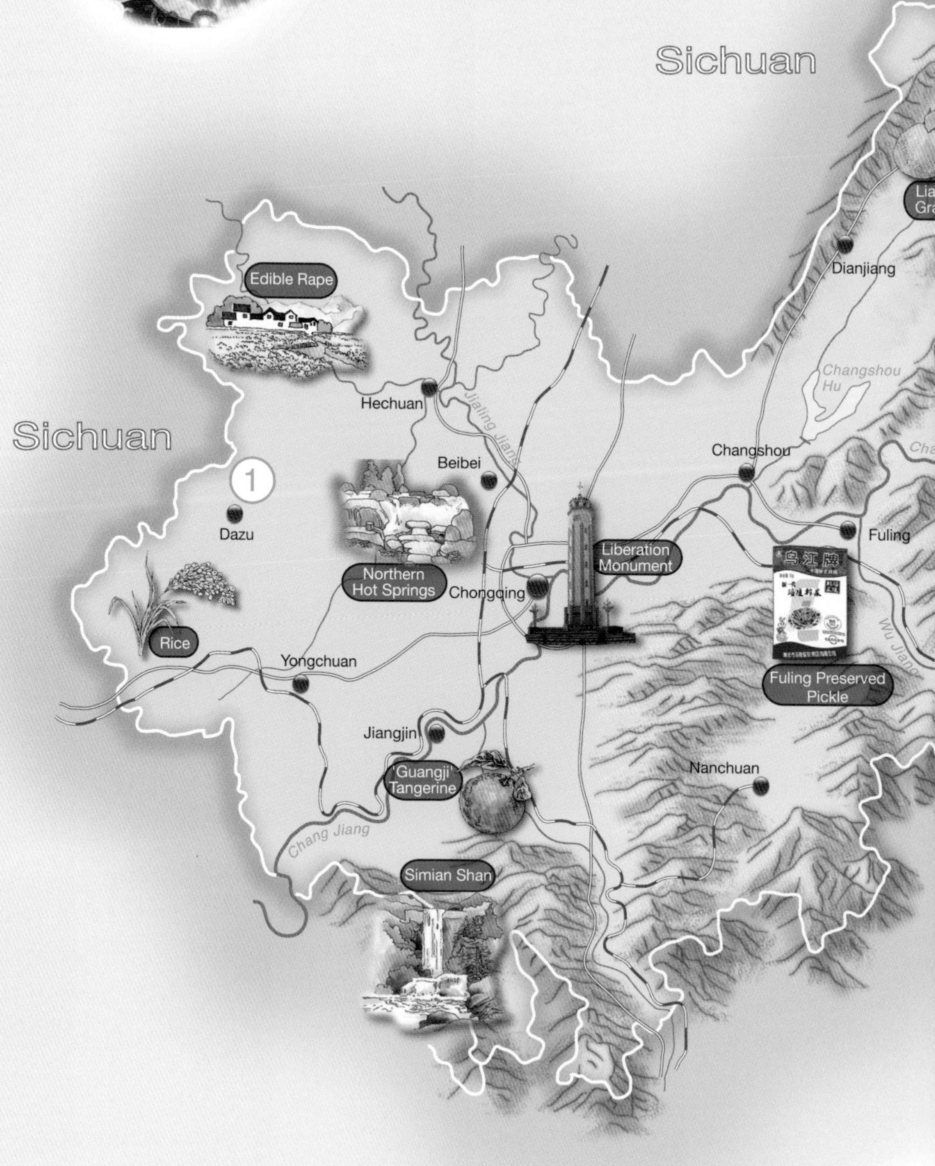

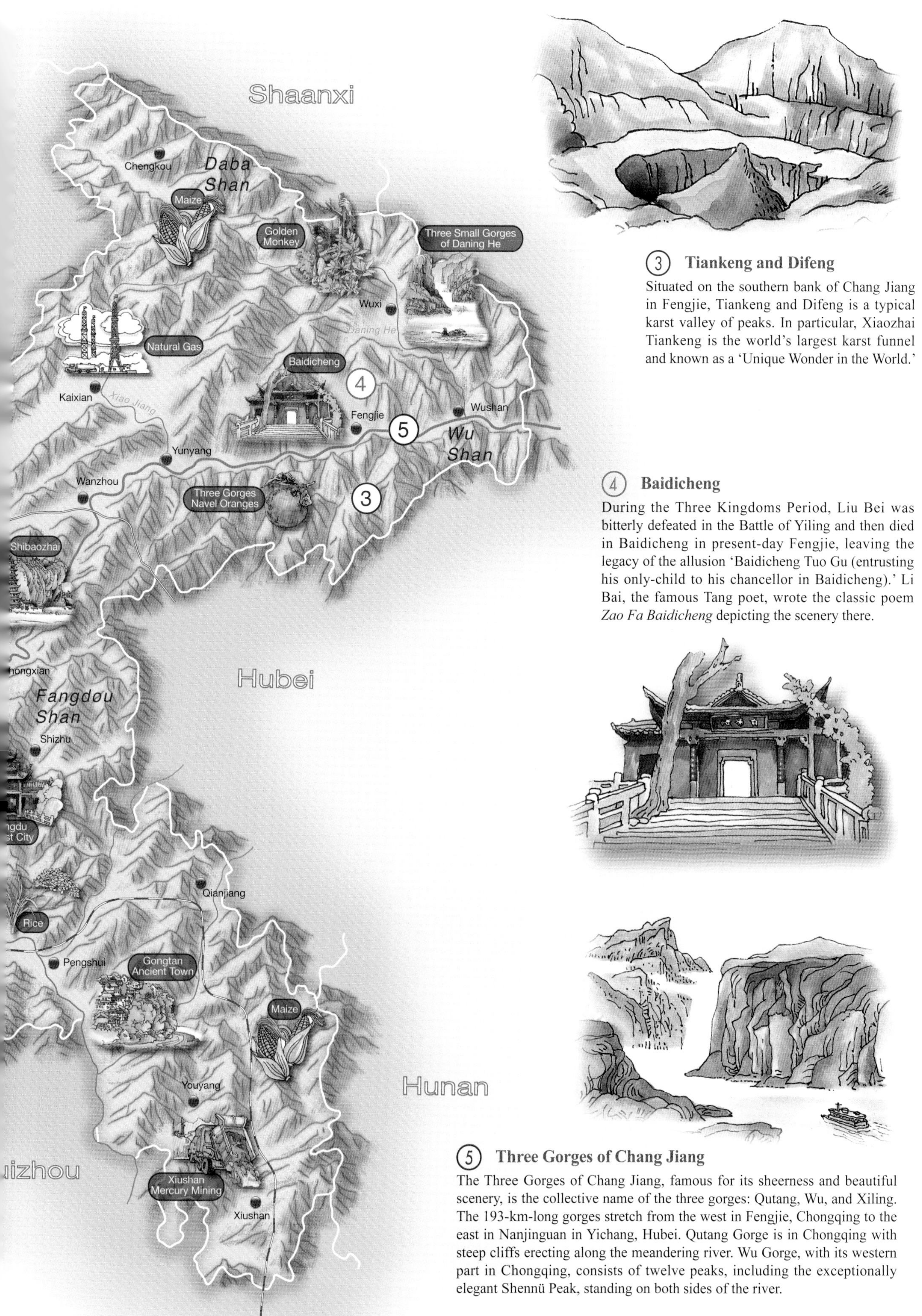

③ Tiankeng and Difeng

Situated on the southern bank of Chang Jiang in Fengjie, Tiankeng and Difeng is a typical karst valley of peaks. In particular, Xiaozhai Tiankeng is the world's largest karst funnel and known as a 'Unique Wonder in the World.'

④ Baidicheng

During the Three Kingdoms Period, Liu Bei was bitterly defeated in the Battle of Yiling and then died in Baidicheng in present-day Fengjie, leaving the legacy of the allusion 'Baidicheng Tuo Gu (entrusting his only-child to his chancellor in Baidicheng).' Li Bai, the famous Tang poet, wrote the classic poem *Zao Fa Baidicheng* depicting the scenery there.

⑤ Three Gorges of Chang Jiang

The Three Gorges of Chang Jiang, famous for its sheerness and beautiful scenery, is the collective name of the three gorges: Qutang, Wu, and Xiling. The 193-km-long gorges stretch from the west in Fengjie, Chongqing to the east in Nanjinguan in Yichang, Hubei. Qutang Gorge is in Chongqing with steep cliffs erecting along the meandering river. Wu Gorge, with its western part in Chongqing, consists of twelve peaks, including the exceptionally elegant Shennü Peak, standing on both sides of the river.

Sichuan Province

In the Song Dynasty, four circuits, Yi, Zi, Li, and Kui, were built and given a collective name: Chuanxia Silu, abbreviated as Sichuanlu, under the control of Sichuan Military Commissioner. That is where the name Sichuan came from. Its abbreviation, Shu, is named after the ancient state of Shu in the area during Shang and Zhou dynasties.

Idiom: Wu Ding Kai Shan

It was a story of the ancient state of Shu. In the Warring States period, the state of Qin rose to power and tried to invade Shu. But Shu was strategically located and the roads were inaccessible for the Qin army to attack Shu. Then the King of Qin ordered to chisel a rock into an ox and put some gold behind it, and spread rumours that the ox could discharge gold. He suggested offering the ox to the King of Shu, who was greedy and immediately ordered five strong men to bring help to pave the road and bring in the ox. When the ox arrived at the capital, the Qin army was right behind it and captured Shu. Since then the road has been called 'Jinniu (Ox of Gold) Road' or 'Shiniu (Ox of Stone) Road' and become a part of the ancient road of Shu. For a long time it has been the essential gate to Shu.

① Huanglong

Located in northern Sichuan, Huanglong is a World Heritage Site as famous as Jiuzhaigou. Besides layers of stunning colourful limestone pools, the limestone formations on the surface of slopes and valleys are like golden dragons wandering through the sea of forests, unfolding an extraordinary, fascinating landscape.

② Dujiangyan Irrigation System

With more than 2,000 years of history, Dujiangyan is a UNESCO World Heritage site and the greatest irrigation project in ancient China. It is in Min Jiang in western Chengdu Pingyuan, and still useful in flood control and irrigation.

③ Jiuzhaigou

Jiuzhaigou County in northern Sichuan is a UNESCO World Heritage site. The extensive rainforest and colourful lakes are as beautiful as a paradise. It is acclaimed as the 'Land of Fairy Tale.'

④ Emei Shan

As one of China's Four Sacred Mountains of Buddhism, it is a well-known tourist attraction and a UNESCO World Heritage Site. The mountains are towering and elegant with luxuriant trees. Sunrise, sea of clouds, Buddha's halo, and sunset are the four great spectacles of Emei Shan.

⑤ Local Snacks of Chengdu

Chengdu is famed as a food heaven with a wide variety of snacks of sensational appearance, aroma, and taste. These snacks are the tourists' favourite delicacies in Chengdu.

⑥ Giant Pandas

Giant pandas are the 'National Treasure' of China that live in the high mountains and valleys near the western border of Sichuan Pendi. Their unique appearance and lovely demeanors win admirers from around the world. However, they are endangered and included in the *UNESCO World Heritage List* in 2006.

Guizhou Province

Named after Gui Shan in the region, Guizhou (abbreviated as Gui) has become the official name of the region from the Yuan Dynasty. It also goes by Qian, as it belonged to Qianzhong Circuit in the Tang Dynasty and its north-eastern part belonged to Qianzhong Prefecture in the Qin Dynasty.

Idiom: Ye Lang Zi Da

Yelang was a small state in the Western Han Dynasty, situated in the west of Guizhou with size approximately equalled a county of Han. In 122 B.C., King Wu of Han was searching for a route to India. He sent envoy to the state of Dian. The King of Dian asked the envoy, 'Which is larger? Han or Dian?' Later the envoy went past Yelang and the King of Yelong asked the same question. Since then 'Ye lang zi da' is used to describe people who are ignorant and self-important.

Idiom: Qian Lü Ji Qiong

There were initially no donkeys in Guizhou. Someone brought in one but found the donkey useless, so it was left at the feet of a mountain. Then a tiger which has never seen a donkey thought it must be ferocious. It went near and tested the donkey. Soon it discovered that the donkey was not fearsome at all and ate it. This is a fable by a Tang literati, Liu Zongyuan and now used to describe people with strong outfit but turn out to be incapable.

① Miao Year

It is the traditional festival of the Miaos to worship their ancestors and celebrate good harvest usually held in the 10th month of the Chinese calendar. During which people wear stunning costume and join the Lusheng Dance, bull fighting, horse racing, and other joyful events.

② Zunyi Conference Site

The Zunyi Conference Site is a two storey structure made of brick and wood which is now turned into the Memorial Hall of the Zunyi Conference. In January of 1935, the Chinese government called up a meeting with significant historical meaning. It paved the way for the completion of the Red Army's Long March of 25,000 li.

③ Maotai — Guizhou's Most Famous Speciality

Maotai is one of the most famous liquors in the world (together with Scotch Whisky and French Brandy).

④ Grey Snub-nosed Monkey of Guizhou

The grey snub-nosed monkey is a rare species found only in China and living in mountains and forests in Fanjing Shan in northeast of Guizhou. It is under the first-grade state protection.

⑤ Huangguoshu Waterfall — China's Largest Waterfall

The Huangguoshu Waterfall is 81 m wide and 74 m high, located in Baishui He in south-western Guizhou.

⑥ Wax Printing

Wax printing is a traditional ethnic dyeing handicraft in south-western Guizhou. It produces clothing with strong ethnical features. The patterns are widely varied; the colours, light and elegant.

① **Stone Forest**

Known as a 'Wonder of the World,' the Stone Forest is a typical karst landscape comprising a set of limestone pinnacles in shape of peaks, pillars, and stalagmites.

Yunnan Province

The name Yunnan is given as the province lies in the 'south of Yun Ling.' During the Qin and Han dynasties, 'Dian' was a large group among ethnic groups in present-day Guizhou, Yunnan, and western Sichuan. They mainly lived in the present-day Dianchi region in Kunming and built their own kingdom. That is why Yunnan is abbreviated as 'Dian.'

Idiom: Zhuang Qiao Wang Dian

The story was recorded in *The Records of the Grand Historian* by Sima Qian. During the reign of Lord Qingxiang of the State of Chu (298—263 B.C.) in the later Warring States period, Zhuang Qiao, the general of Chu, was appointed to lead troops to the south and expand the territory of Chu. He kept on fighting and reached Dianchi. At that time Ba and Qianzhong prefectures of Chu were occupied by the State of Qin, meaning that Zhuang and his army had no way to return to Chu. Therefore, he stayed in Dianchi where he built the Dian Kingdom and proclaimed himself the King of Dian (or 'King of Zhuang').

Idiom: Qi Qin Meng Huo

In 225 A.D., Zhuge Liang, the chancellor of the State of Shu, led the Shu army to settle the rebellion in Nanzhong region. Nanzhong in present-day Yunnan was the home of minority groups. The phrase 'Qi Qin Qi Zong' is a psychological tactic by Zhuge Liang to capture Meng Huo, a powerful local leader of minority groups.

② **Old Town of Lijiang**

With over 800 years of history, Lijiang is a delicate, primitively simple city of Naxi ethnic group. It is a UNESCO World Heritage site and reputed as the most beautiful ancient city in China and the 'Oriental Venice.'

③ **Water Splashing Festival of Dai Ethnic Group**

Water Splashing Festival is a traditional event held annually in mid-April. During the event, people splash water on one another as a blessing and for celebrating the New Year of the Dai calendar.

④ **World Expo Garden in Kunming**

The World Horticultural Exposition Garden in Kunming was the venue for the World Horticultural Exposition in 1999.

⑤ Xishuangbanna — Home to Peacocks

Xishuangbanna is a beautiful, bizarre land with abundant resources. It is a natural kingdom of animals and plants as its tangled tropical rainforest is the habitat of many rare animals such as peacocks and Asian elephants. The icon of a golden peacock is commonly used to represent Yunnan.

⑦ Yuanmou Man

Yunnan is one of the important birthplaces of human civilisation. The Yuanmou Man from Yuanmou of Yunnan 1.7 million years ago was the first ape men in Asia.

⑥ Three Pagodas of Dali

Dali is a famous city of rich history and culture in China. It is at the feet of Cang Shan beside Er Hai. Built in the Tang Dynasty, the three pagodas has been the icon of the ancient city of Dali.

Xizang Zizhiqu

Xizang Zizhiqu (Tibet Autonomous Region) was called Wusi Zang in the Ming Dynasty and Wei Zang in Qing. Wei means the front part (Lhasa and Shannan) and Zang means the rear part (Xigazê and Gyangzê). During the reign of Kangxi of Qing, the Emperor changed it into Xizang, which has been adopted ever since.

Receiving the Tibetan Envoy

The *Emperor Taizong Receiving the Tibetan Envoy* was drawn by the Tang painter Yan Liben, depicting the scene of Emperor Taizong of Tang receiving the Tubo envoy Lu Dongzan. At the beginning of 7th century, the Tubo kingdom became stronger and conquered Qingzang Gaoyuan under the leadership of the Tubo King, Songtsen Gampo. In 640 A.D. (Tang), he sent a special envoy, Lu Dongzan, to Tang to ask for marriage alliance. Taizong agreed and let Princess Wencheng, a member of the royal family, to marry him. Thanks to the marriage, Princess Wencheng brought in the culture of the Central Plain to

Tubo. The marriage has a profound and far-reaching influence on developing friendship between Han and Tibetan ethnic group, fostering economic and cultural communication and development between the Central Plain and the Tubo region. For example, the Ramoche Temple in Lhasa was built when Princess Wencheng went into Tibet.

① Potala Palace

Construction of the Potala Palace started in 7th century during the reign of Songtsen Gampo. It was reconstructed in 17th century and since then became the winter palace of generations of Dalai Lamas. It is not only the world's largest collection of palaces, but also an icon of Lhasa. It was enlisted as a UNESCO World Heritage site in 1994.

② Tibetan Pilgrims

In Tibet, you can always see pilgrims coming from afar to Lhasa. They arrive in Tibet without any transport but only by themselves. They stop every few steps to prostrate themselves and continue their journey piously towards the sacred city of Lhasa.

③ Qomolangma Feng

Qomolangma Feng is at the boundary of China and Nepal with an elevation of 8,844.43 m. It is the world's highest peak and called the 'Top of the World.'

④ Qingzang Railway

The railway started from the east in Xining of Qinghai to the west in Lhasa of Tibet and began operation on 1st July 2006. It is the world's highest and longest highland railway, as well as the first ecologically friendly railway on a snowy plateau.

⑤ Food and Drinks of Tibetan Ethnic Group

Typical Tibetan food include highland barley wine, Buttered tea, Tsamba (paste of cooked, glutinous rice).

⑥ Yarlung Zangbo Grand Canyon

Located in the lower course of Yarlung Zangbo Jiang, the Yarlung Zangbo Grand Canyon is the longest and deepest canyon in the world.

Shaanxi Province

In the early years of Western Zhou Dynasty, Shaanyuan (present-day Shaanxian) was divided into the east, called 'Shaandong' ruled by Zhou Gong, and the west, 'Shaanxi' ruled by Zhao Gong. That is the origin of the name Shaanxi.

Starting Point of the Silk Road on Land

The road was formed in the Western Han Dynasty when the envoy Zhang Qian was assigned to explore the Western Regions. It started from Chang'an (present-day Xi'an) through Gansu, Xinjiang, Central Asia and Western Asia, linking to channels in various countries in the Mediterranean Sea. It was named 'Silk' as silk was the most important good transported to the west. In Sui and Tang dynasties, the Silk Road went into unprecedented prosperity. Foreign merchants were commonly seen in Chang'an. In mid-Tang, the Silk Road on land was disrupted by the frequent wars and later replaced by the Silk Road on the sea.

Idiom: Jing Wei Fen Ming

Jing He is the largest branch of Wei He. When the two rivers converge in the countryside in northern Xi'an, they form a stark contrast that one of them is clear while another is obscure due to different amount of sand. As it is bizarre that two kinds of water flowing in the same riverbed, people use it as a metaphor for a clear boundary, the difference between right and wrong, good or bad people, or opposite opinions of the same issue.

① Yaodong in Northern Shaanxi

Yaodong are cave houses of residents on the Huangtu Gaoyuan. Typical examples are the ones in northern Shaanxi. Most of them are dug and built along the slopes. They are strong, resistant, cooler in summer while warmer in winter.

② Terracotta Army

The Terracotta Army is the ancillary structure to the Qin Emperor's tomb. The enormous size and excellent artistic level of the warrior and horse statues awe the world.

③ Huang Di Mausoleum

Huang Di Mausoleum is the tomb of Huang Di, named Xuan Yuan, the common ancestor of Chinese. The tomb is on Qiao Hill in Huangling County, where Chinese home and abroad come and worship him.

④ Shaanxi History Museum

Shaanxi History Museum is a large national museum with modern facilities, situated in the north-western side of Dayan Tower in Xi'an. It contains over 370,000 precious items and thus called the 'Treasure of the Chinese Culture.'

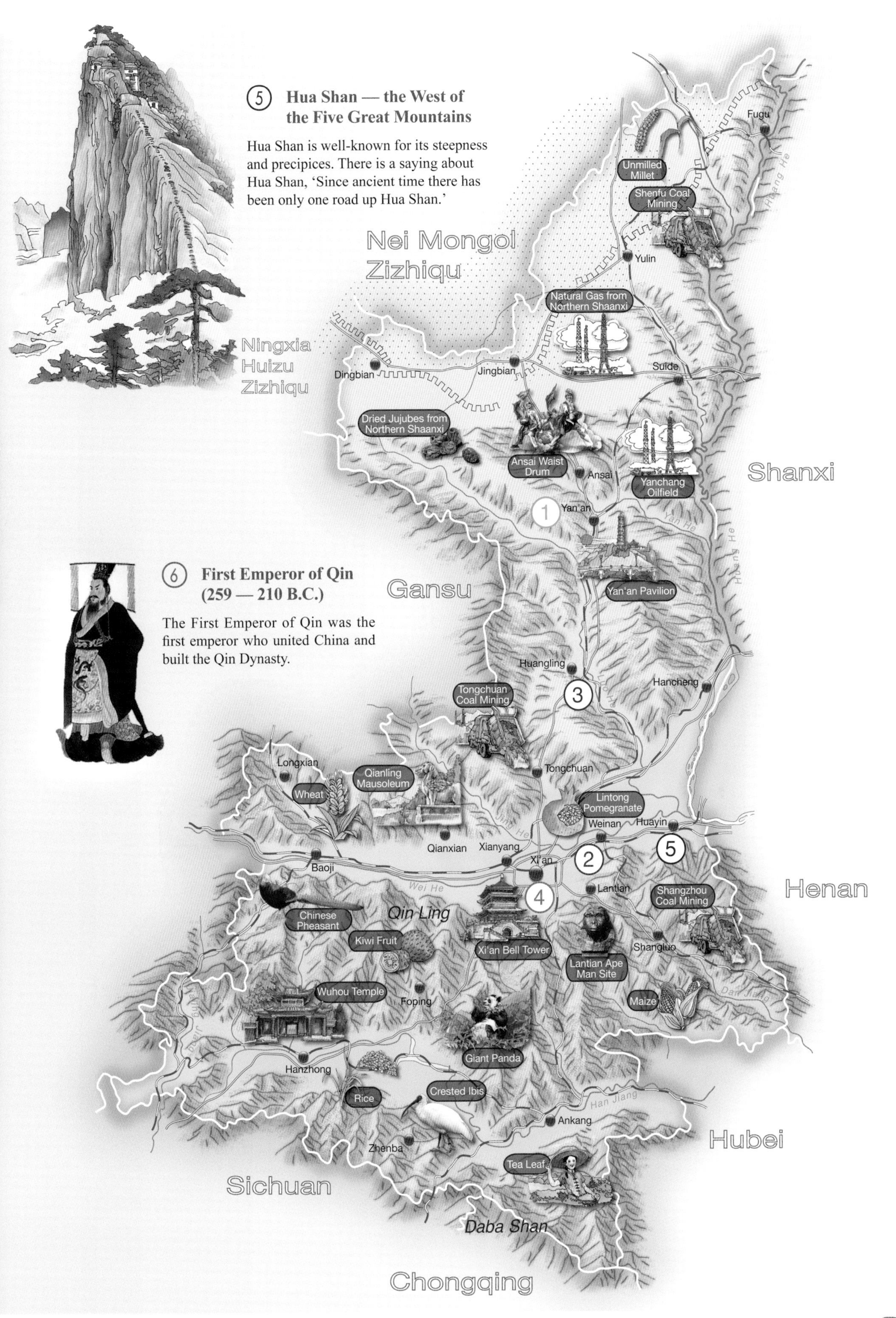

⑤ Hua Shan — the West of the Five Great Mountains

Hua Shan is well-known for its steepness and precipices. There is a saying about Hua Shan, 'Since ancient time there has been only one road up Hua Shan.'

⑥ First Emperor of Qin (259 — 210 B.C.)

The First Emperor of Qin was the first emperor who united China and built the Qin Dynasty.

MONGOLIA
Xinjiang Uygur Zizhiqu
Bei Shan
Cotton
Yumen Oilfield
Guazhou
Yumen
Dunhuang
Shule He
Jiayuguan
Qilian Shan
Dang He
Danghe Nanshan

Gansu Province

Gansu is named after Ganzhou (present-day Zhangye) and Suzhou (present-day Jiuquan) in the region. As most of its land is in the west of Long Shan (present-day Liupen Shan), a circuit name 'Long You Dao,' meaning the right of Long, was set up in the Tang Dynasty. That's why 'Long' has become the abbreviation of the region. In the Western Xia Dynasty, it was named 'Gan' as its administrative centre was in Ganzhou.

Yumenguan and Yangguan

Yumenguan and Yangguan are in Gansu and are mentioned in two classic Tang poems written by Wang Zhihuan and Wang Wei respectively.

Yumenguan was built during the reign of Emperor Wu of Han when he ordered to open up routes to the Western Regions and set up four prefectures of Hexi. 'Yumen,' meaning the gate to jade, was named so because of the import of jade through the pass. It was the gate to various destinations in the Western Regions in the Han Dynasty.

Yangguan was a vital transport point on land in ancient China and the essential gate to the Southern Silk Road. It was set up in the Western Han Dynasty. The pass, situated in the south of Yumenguan, was named the Yangguan as ancient Chinese called the north as Yin and south as Yang. In the Tang Dynasty when the eminent monk, Xuanzhuang, returned from his study of Buddhism in India, he took the Southern Silk Road through the Yangguan and went back to Chang'an (capital of the Tang Dynasty).

② Jiayuguan

Jiayuguan is a well conserved pass and the westernmost point of the Great Wall. It was an important place of military due to its strategic location.

Yumenguan

Yangguan

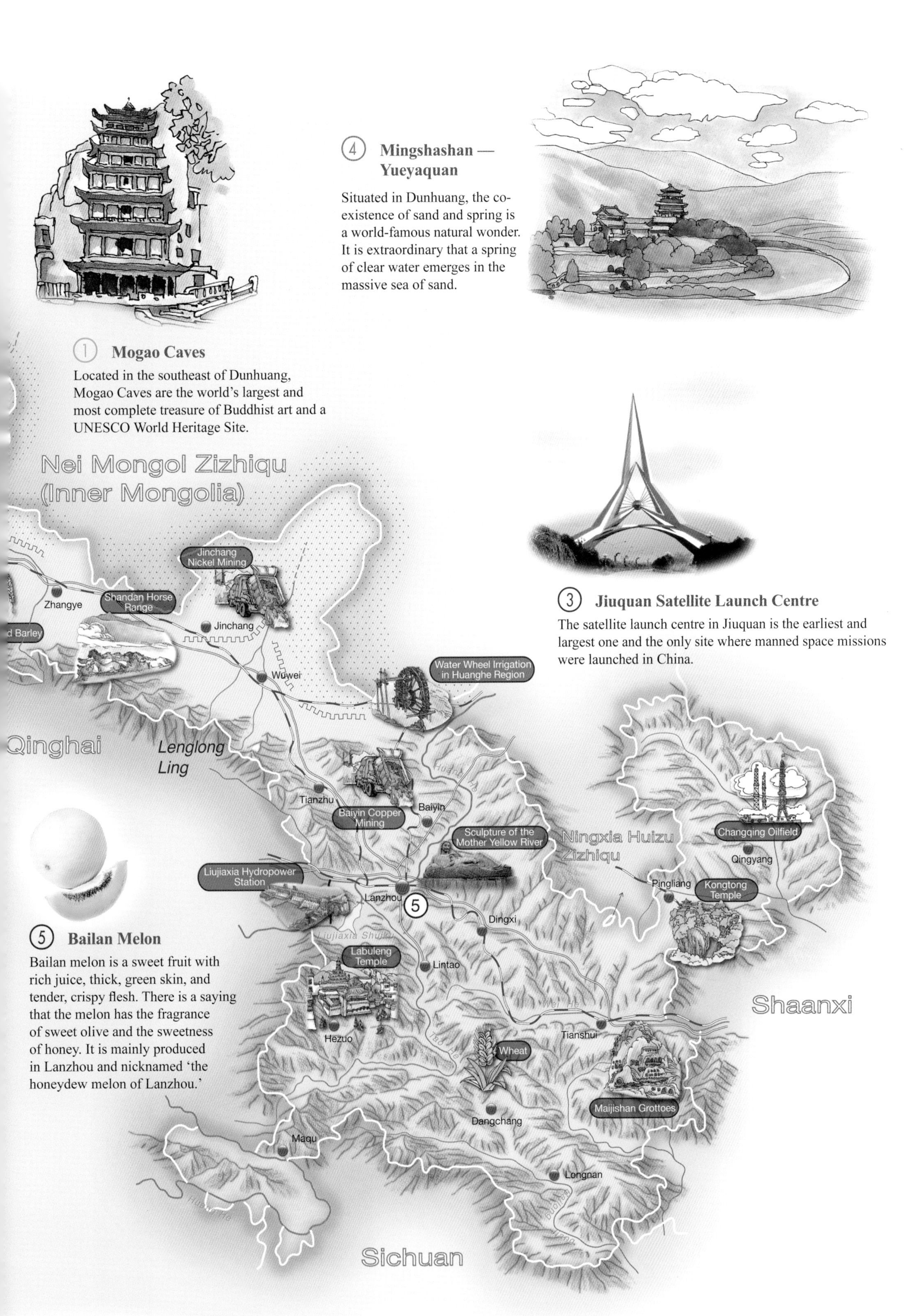

④ **Mingshashan — Yueyaquan**

Situated in Dunhuang, the co-existence of sand and spring is a world-famous natural wonder. It is extraordinary that a spring of clear water emerges in the massive sea of sand.

① **Mogao Caves**

Located in the southeast of Dunhuang, Mogao Caves are the world's largest and most complete treasure of Buddhist art and a UNESCO World Heritage Site.

③ **Jiuquan Satellite Launch Centre**

The satellite launch centre in Jiuquan is the earliest and largest one and the only site where manned space missions were launched in China.

⑤ **Bailan Melon**

Bailan melon is a sweet fruit with rich juice, thick, green skin, and tender, crispy flesh. There is a saying that the melon has the fragrance of sweet olive and the sweetness of honey. It is mainly produced in Lanzhou and nicknamed 'the honeydew melon of Lanzhou.'

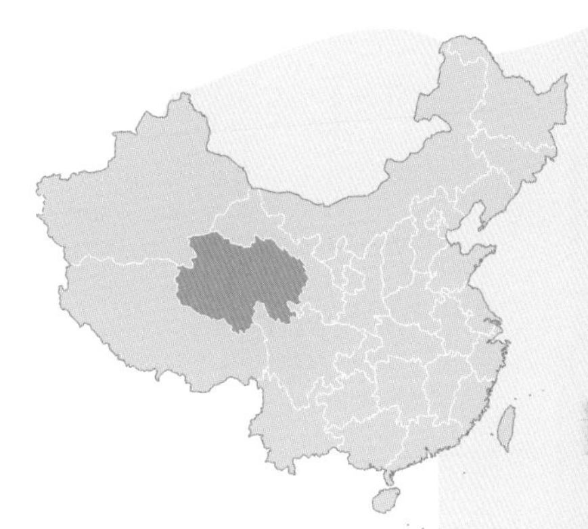

Qinghai Province

The name Qinghai is named after Qinghai Hu, the largest inland salt lake in China. Qinghai is called Cuo'erbo (Tibetan) or Kuku Nur (Mongolian) meaning 'Green Lake.'

Qinghai — the Bleak, Snowy Battlefield

Qinghai appears in the Tang poem *Congjun Xing* in which the bleak scenery of a snowy Qinghai under long stretches of clouds is depicted by Wang Chanling. In the Tang Dynasty, Qinghai was the main battlefield of Tang and its major enemies, Tubo in the west and Turks in the north. The famous general Geshu Han built a city here and installed the Shenwei army as strong defence of the border against foreign enemies.

Wang also mentioned in *Congjun Xing* that the army in Qinghai fought in the north of Tao He and finally captured Tuyuhun. Tuyuhun was a branch of the Xianbei ethnic groups living in northwestern China such as Qinghai and Gansu. When it was conquered by Tang in early Tang Dynasty, its leader was titled the King of Qinghai. Tao He, originating from Xiqing Shan in Qinghai, is a branch of the upper course of Huang He.

① Origin of Huang He

The headwaters of the upper course of Huang He consist of Kar Qu, Yoigilonglêb Qu, Za Qu, and several lakes such as Ngoring Hu and Gyaring Hu.

② Tibetan Antelope — 'Fairies of the Plateau'

Tibetan antelopes mainly live on Qingzang Gaoyuan. They are good at running and under the first-grade state protection.

③ Hoh Xil

Hoh Xil is China's largest no-man's land in western Qinghai. Despite the poor natural resources, Hoh Xil is home to many rare wildlife species such as Tibetan antelopes, kiangs, yaks, and snow leopards.

④ Origin of Chang Jiang

The headwaters of Chang Jiang consist of the official headstream Tuotuo He, the southern headstream Dam Qu, and the northern headstream Qumar He. The origin of Chang Jiang is part of the Sanjiangyuan National Nature Reserve (which also includes the origins of Huang He and Lancang Jiang).

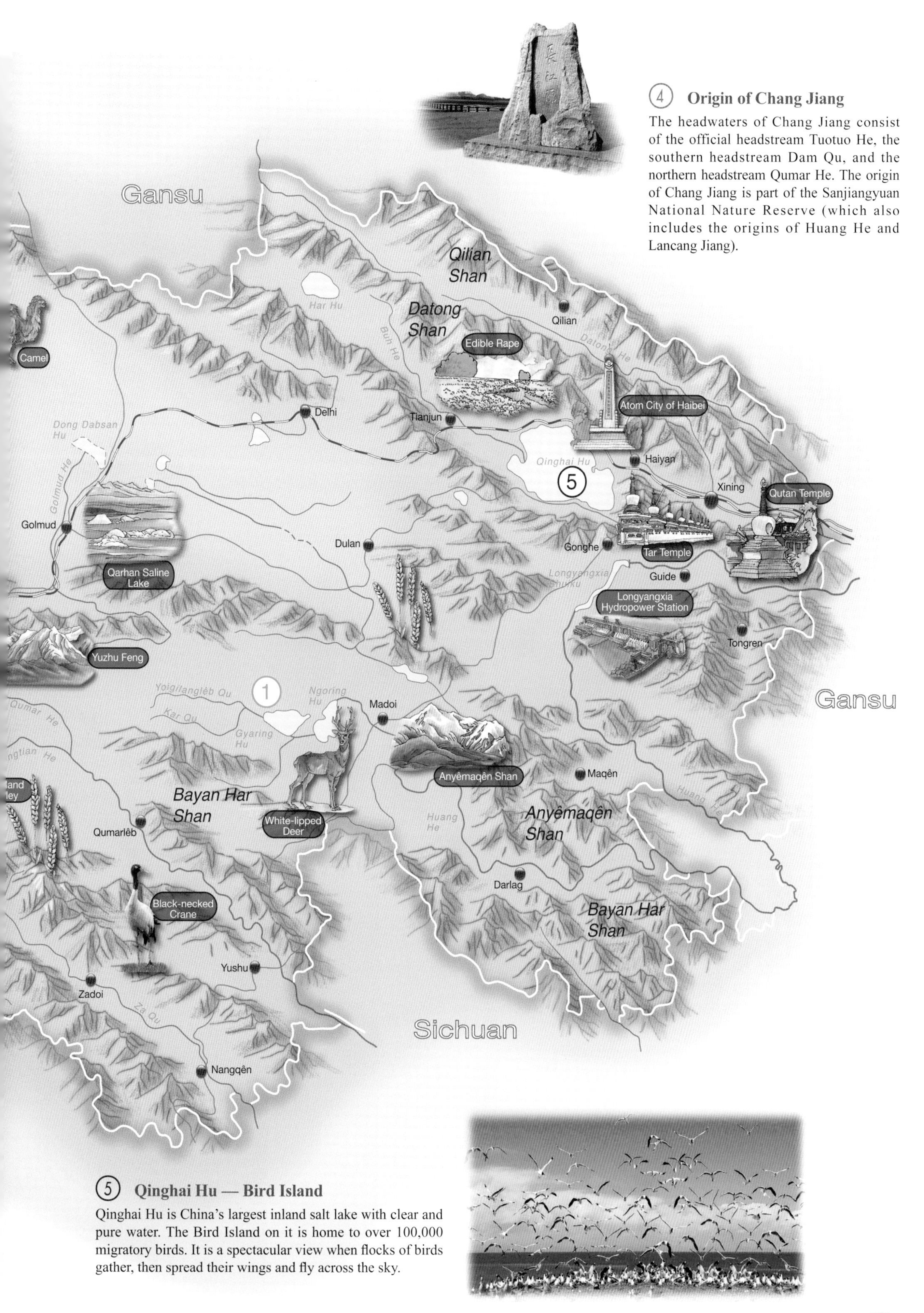

⑤ Qinghai Hu — Bird Island

Qinghai Hu is China's largest inland salt lake with clear and pure water. The Bird Island on it is home to over 100,000 migratory birds. It is a spectacular view when flocks of birds gather, then spread their wings and fly across the sky.

Ningxia Huizu Zizhiqu

Ningxia was part of the Western Xia Kingdom in the final years of the Song Dynasty. After ending the Western Xia Dynasty, rulers of the Yuan Dynasty set up Western Xia Province there, then renamed it as Ningxia Province, implying a wish for peace in Western Xia. That is the origin of the name Ningxia.

Ningxia and the Western Xia kingdom (1038—1227)

In the Northern Song Dynasty, Li Yuanhao of the Dangxiang group built the Da Xia Kingdom, more commonly called Western Xia Kingdom, with Yinchuan as the capital and a territory covering most of the area of present-day Ningxia, Shanxi, and Gansu. It was ended by the Yuan Dynasty.

Dangxiang is a branch of the Qiang ethnic group (other sources suggest it was from the Xianbei-Tuoba tribe). During the Tang Dynasty, they lived in Xiazhou, Youzhou (both in present-day Ningxia). During Huangchao Rebellion in the final years of the Tang Dynasty, Tuoba Sigong, the leader of Xianbei group sent armies to invade the Tang Empire. After the war, he was bestowed by Emperor Xizong of Tang the title of Military Commissioner of Xiazhou and the Duke of Kingdom of Xia, and the surname of Li. His bestowal established the governing territory of his clan and paved the way for the rise of the Western Xia Kingdom.

① Sheepskin Raft

Sheepskin rafting is an old ferry across Huang He, typically made by around a dozen parallel sheepskin rafts attached to a wooden frame. The raft is buoyant, light, and easy to operate.

② Lesser Bairam (Fast-breaking Festival)

Held at the start of the 10th month of the Islamic calendar, the Lesser Bairam is a grand event of Muslim ethnic groups such as the Hui people. On that day, people need to clean their body and dress in new clothes, then gather in mosques nearby to attend ceremonies and other solemn festive activities.

③ Shahu Nature Reserve

Situated in the north of Yinchuan, Shahu Nature Reserve is a scenic area combining the scenery of Jiangnan watertowns and a vast desert.

④ Western Xia Mausoleums

Western Xia Mausoleums are in the vast expanse of Gobi Desert at the feet of the east of Helan Shan. They are the imperial tombs of the Western Xia Dynasty and known as the 'Oriental Pyramid.'

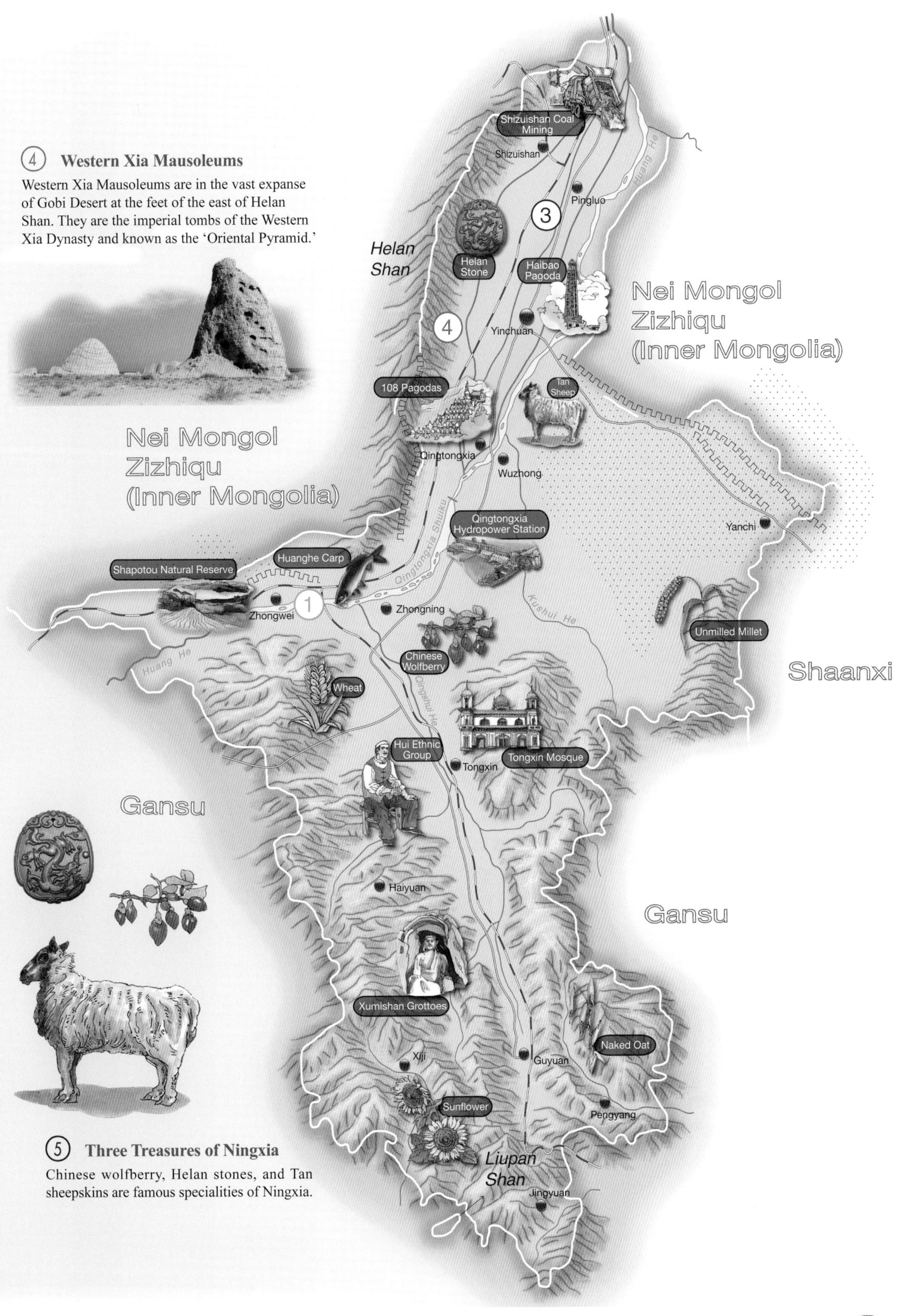

⑤ Three Treasures of Ningxia

Chinese wolfberry, Helan stones, and Tan sheepskins are famous specialities of Ningxia.

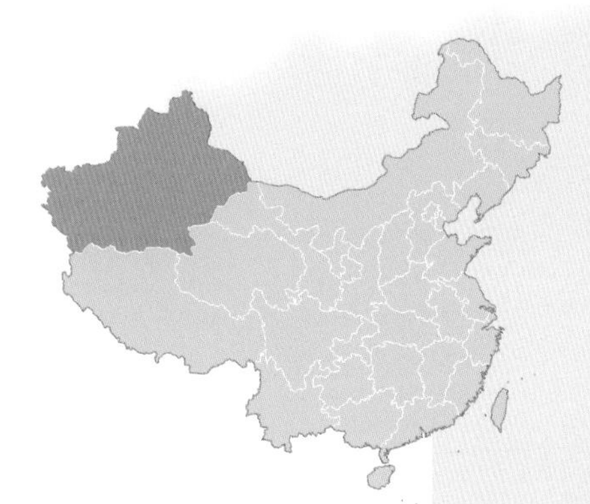

Xinjiang Uygur Zizhiqu

Xinjiang is part of the Western Regions in ancient China. In the Han Dynasty, Protector-General of the Western Regions was established here. In the Tang Dynasty, the Anxi Protectorate and Beiting Protectorate and other official departments were also set up here. In 1757 after the Pacification of the Dzungars in the region, it was designated by Qianlong as 'Xinjiang,' meaning 'an old territory returned to the motherland.' In 1884, the Qing government established Xinjiang Prefecture here.

Snowy Landscape of Tian Shan

The Tang poet Cen Shen depicted the snowy sights of Tian Shan in *Baixue Ge Song Wu Panguan Guijing*, a poem about seeing his friend off in a place called Luntai. He likened the snow-covered trees and Tian Shan to blossoms of a thousand pear trees. Luntai in the Tang Dynasty is the present-day Ürümqi, the capital of Xinjiang. There was another Luntai in the Han Dynasty, which is the present-day Luntai County in Xinjiang, where Emperor Wu of Han sent troops to open up new land.

① Tian Chi in Tian Shan

Tian Chi, known as 'Jade Pool' in ancient times, is on the midway to Bogda Peak of Tian Shan. The scenery is as beautiful as paradise with emerald water reflecting snowy peaks surrounded by Chinese spruces.

② Ancient City of Loulan

Loulan was a well-off ancient city on Silk Road in southern Xinjiang. It was once very prosperous but now it is only some ruins for memorial.

③ Xinjiang International Grand Bazaar

Bazaar is a Uygur phrase meaning 'markets.' The International Grand Bazaar is a shopping centre with distinctive features of the Western Regions and the first choice for shoppers.

④ Ranch in Tian Shan

Rivers originated from Tian Shan form expansive oases which become a natural ranch.

⑤ Home of Fruits

Xinjiang has long been reputed as the 'Home of Fruits,' producing grapes, melons, pears, and pomegranates. A large variety of quality fruits, both fresh and dried, can be seen in markets throughout the year.

⑥ Grape Valley — Singing and Dancing of Uygur Ethnic Group

Grape Valley is in the northeast of Turpan and the west of Huoyan Shan and famous for producing grapes. Now it is developed into a large tourist area combining agricultural sightseeing, catering, and the singing and dancing of Uygurs. Uygurs, one of the 56 ethnic groups in China, are talented singers and dancers. They mainly live in the south of Tian Shan.

Hong Kong Special Administrative Region

Hong Kong was a natural harbour near a stream of pure and sweet water. Sailors often stopped at Hong Kong to drink water. As time went by, the reputation spread and the place was named 'Heung Gong' meaning 'fragrant stream.' Hence the harbour formed by the sediments brought by the 'fragrant stream' when it flows into the sea is called 'Hong Kong.'

Unequal Treaties that Separated Hong Kong from China

Hong Kong comprises three parts: Hong Kong Island, Kowloon Peninsula, and the New Territories. The three regions originate from three unequal treaties. During the First Opium War in 1840, the British government forced the Qing government to sign the Treaty of Nanjing (also known as the Treaty of Jiangning) in 1842 and surrender Hong Kong Island.

In 1856, the Anglo-French army started the Second Opium War and forced the Qing government to sign the Treaty of Peking in 1860 and surrender Kowloon Peninsula, that is, the region in the south of the present-day Boundary Street.

In 1898, the British government forced the Qing government to sign the Convention for the Extension of Hong Kong Territory and to offer a 99-year lease on New Territories until 30 June 1997.

① Handover of Hong Kong

From 1982 to 1984, China and Great Britain discussed the future of Hong Kong. In 1984, they signed the China–United Kingdom Joint Declaration on the Question of Hong Kong and concluded that the People's Republic of China would resume its right to rule over Hong Kong on 1 July 1997.

② Ocean Park Hong Kong

Ocean Park is a renowned Asian marine park in southern Hong Kong Island. With an expansive aquarium, the park is a popular attraction for fun and entertainment.

③ HKSAR — regional flags and emblem

The regional flag of HKSAR is red with a white Bauhinia flower in the middle. The red flag symbolizes China and the white flower represents Hong Kong. The emblem is round with a central pattern of the same symbolic meaning as that in the flag. The outer band shows the full name of 'HKSAR of the PRC' in Chinese and 'Hong Kong' in English.

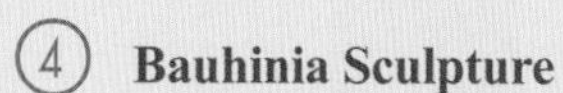

④ Bauhinia Sculpture

Bauhinia is the flower of Hong Kong. The golden sculpture is given by the Central Government of China, now displayed in the Golden Bauhinia Square as a memorial of Hong Kong's return to the mother country.

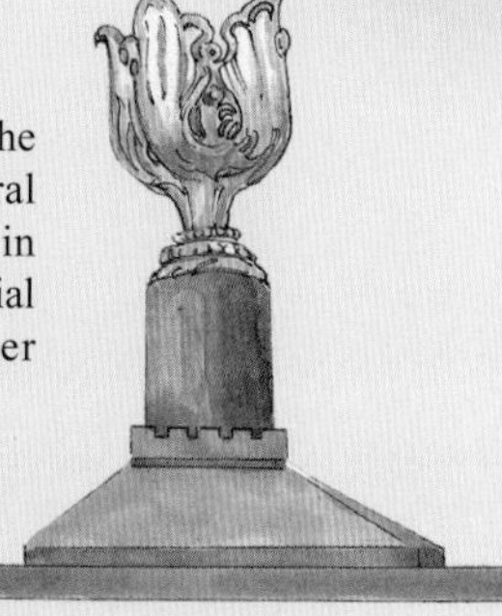

⑤ Horse Racing

The Hong Kong Jockey Club is the world's largest horse racing organisation with a well-known racecourse in Sha Tin.

Lantau Channel

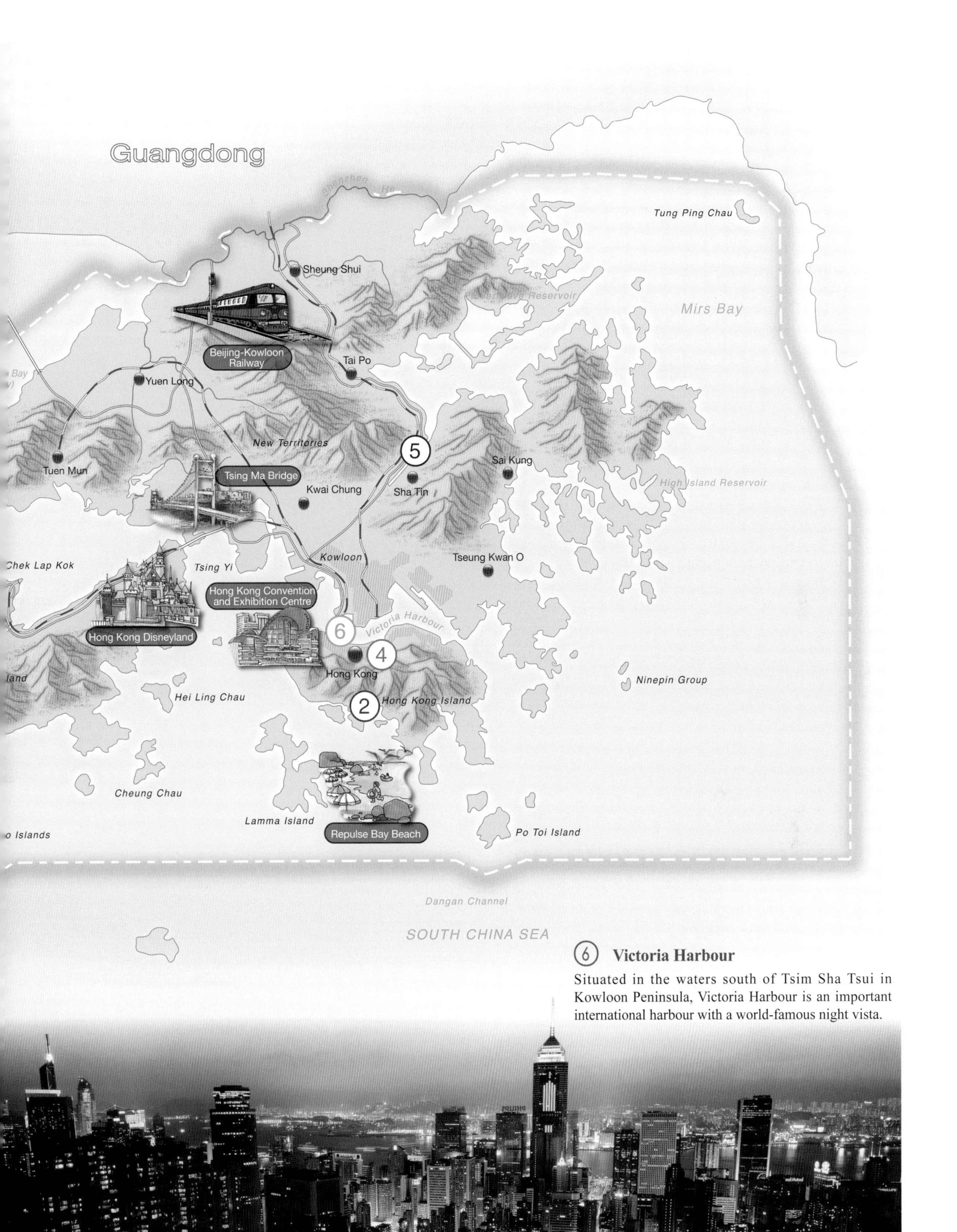

⑥ Victoria Harbour

Situated in the waters south of Tsim Sha Tsui in Kowloon Peninsula, Victoria Harbour is an important international harbour with a world-famous night vista.

Macau Special Administrative Region

Macau used to be a small fishing village named Hou Keng or Hou Keng Ou. At that time, a port could be called 'Ou,' so Macau as a port is called 'Ou Mun.'

Macau occupied by Portuguese

In 1557, Portuguese obtained the right of abode in Macau from the provincial government of Guangdong in the Ming Dynasty. They became the first Europeans to enter China. When they landed on Macau near A-ma Temple, they asked the locals the name of this place. The locals answered 'A Ma Gao,' so the Portuguese named it Macau. On 1 December 1887, the Qing government signed the Lisbon Protocol and the Treaty of Friendship and Trade with Portugal. The treaties became the diplomatic means by Portugal to officially occupy Macau.

Destination for Priests

Since Portuguese gained the right to live in Macau, priests from Europe came to Macau for easier entry into China. On 7th August 1582, Matteo Ricci (1552—1610), the Italian Jesuit priest, arrived in Macau. He first spread Christianity in Zhaoqing and Nanjing. It was until 1601 that he was permitted to meet Emperor Qianlong and to live regularly in Beijing. In 1610, he passed away due to illness and was buried in Beijing.

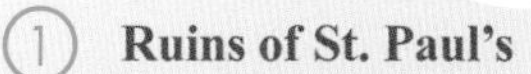

① Ruins of St. Paul's

The remains of the facade of the Church of Mater Dei is now a landmark of Macau. The historic centre, where the facade being the best-known attraction, is the most complete and best-preserved collection of architecture in China showing co-existence of Chinese and Western features. It is also a UNESCO World Heritage site.

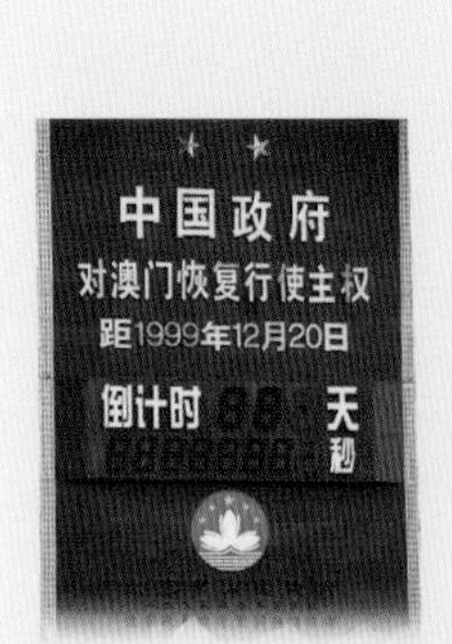

② Macau's Return to China

In 1986, Chinese and Portuguese government started negotiating the future of Macau. In 1987, the premiers of both countries signed the 'Joint Declaration on the Question of Macau' and the annexes. On 20 December 1999, China resumed its right to rule in Macau.

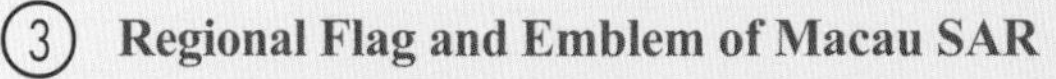

③ Regional Flag and Emblem of Macau SAR

The regional flag of Macau is green with 5 stars, a lotus flower, a bridge and sea water, implying that Macau is an inseparable part of China and wishing Macau success and affluence. The emblem is round with the same pattern in the centre and same implication as those in the flag. The outer band shows the full name of 'Macau SAR of the PRC' in Chinese and 'Macau' in Portuguese.

④ Hotel Lisboa

Located in the Macau Peninsula, Hotel Lisboa is the biggest world-class five-star hotel complex in Macau combining accommodation, entertainment, and gambling. It is particularly well-known for having the largest casino in Asia.

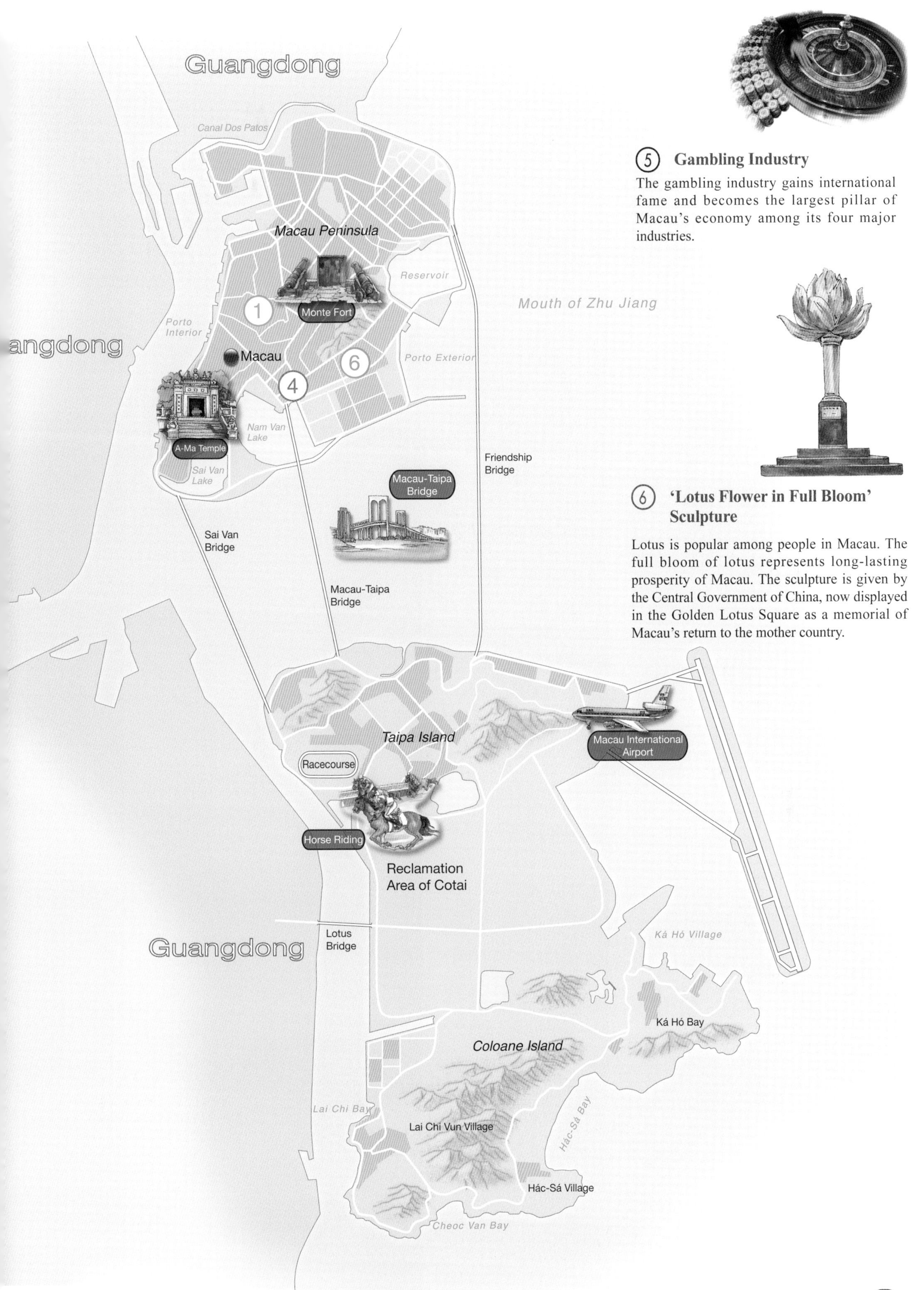

⑤ Gambling Industry

The gambling industry gains international fame and becomes the largest pillar of Macau's economy among its four major industries.

⑥ 'Lotus Flower in Full Bloom' Sculpture

Lotus is popular among people in Macau. The full bloom of lotus represents long-lasting prosperity of Macau. The sculpture is given by the Central Government of China, now displayed in the Golden Lotus Square as a memorial of Macau's return to the mother country.

Taiwan Province

Taiwan had nearly ten names throughout history, such as Yingzhou, and Yizhou. The name 'Taiwan' first appeared in a government notice during the reign of Wanli of Ming over 300 years ago. That was the first appearance of Taiwan in an official, traceable historical resource. In 1885, the Qing government officially set up the Taiwan Province.

Restoration of Taiwan by Zheng Chenggong

In the first half of 17th century, Taiwan gradually became a Dutch colony. In April of 1661, Zheng led an army of 25,000 and several hundreds of battleships to Taiwan from Jinmen. After fierce battles, he succeeded in restoring Taiwan from the Dutch in February of 1662.

Treaty of Maguan and Surrender of Taiwan

In 1894, Japan started the First Sino-Japanese War. The Qing government was defeated in the following year and forced to sign the humiliating Treaty of Maguan on 17 April, ceding Taiwan and its islands to Japan. Since then Taiwan has become a Japanese colony for 50 years. Taiwan finally returned to the territory of China after the victory of the Second Sino-Japanese War.

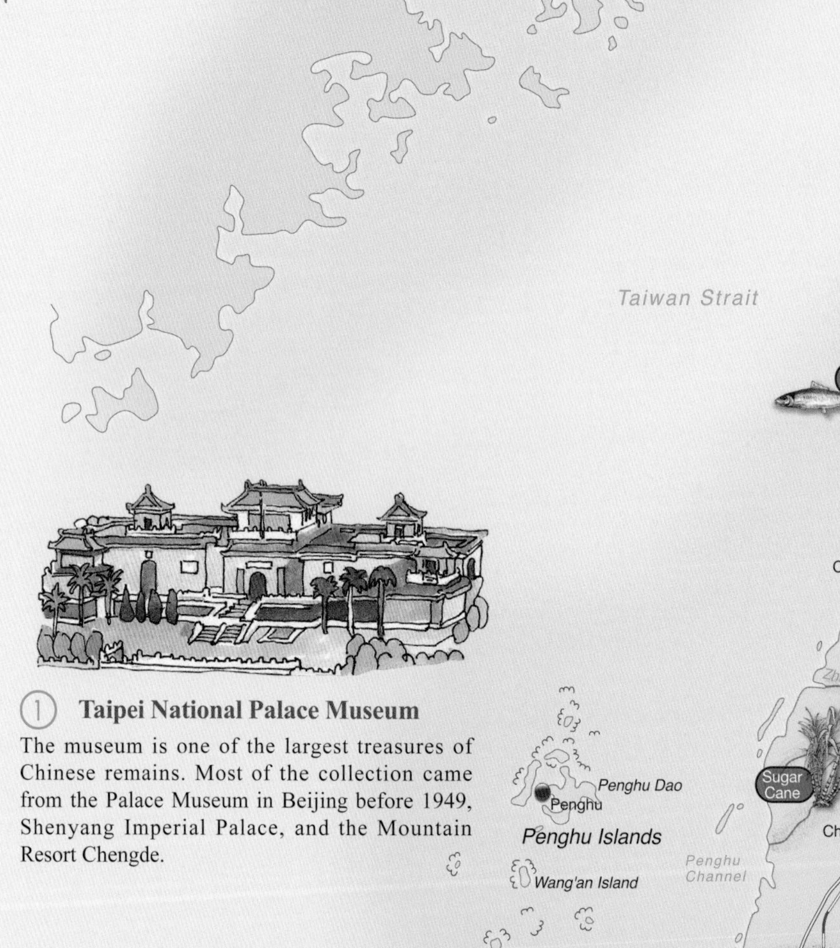

① Taipei National Palace Museum

The museum is one of the largest treasures of Chinese remains. Most of the collection came from the Palace Museum in Beijing before 1949, Shenyang Imperial Palace, and the Mountain Resort Chengde.

② Alishan — Symbol of Taiwan

Situated in northeast of Jiayi County, Alishan is most famous for its three natural wonders: forests, sea of clouds, and sunrise.

Huangwei Yu

Chiwei Yu

Diaoyu Dao

EAST CHINA SEA

Pengchia Yu

Mienhua Yu

Jilong Gang

Jilong

Taoyuan

Tamsui River

Taipei

Yilan

Areca Nut

Tuna

Sugar Cane

Taroko National Park

Hualian

Pineapple

Taiwan Dao

Coral

PACIFIC OCEAN

Gaoshan Ethnic Group

Lobster

Taidong

Lu Dao

Offshore Fishing Industry

Lan Yu

③ Sun Moon Lake

Located in Nantou County, Sun Moon Lake is the only natural lake in Taiwan and a scenic area for tourism. A small island in the lake divides it into two parts: one is round shaped like the sun, another is curved shaped like the crescent, hence the name.

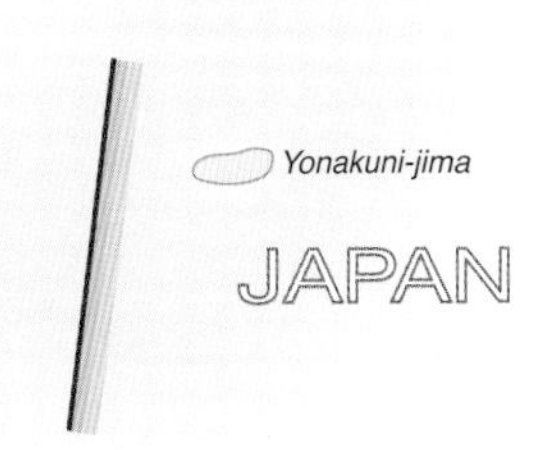

④ Gaoshan Ethnic Group

Gaoshan people are one of the minority groups in Taiwan Dao. They usually live in the mountainous areas in central Taiwan and rift valley plains in eastern Taiwan. They are good at singing and dancing. Their pestle dance and long-hair dance are very artistic.

⑤ Taiwan Specialities — Orchids, Butterflies

Taiwan is known as the 'Treasure Island' for its beauty and plentiful resources. Fruits abound in Taiwan which is also a world-famous habitat of orchids and butterflies.

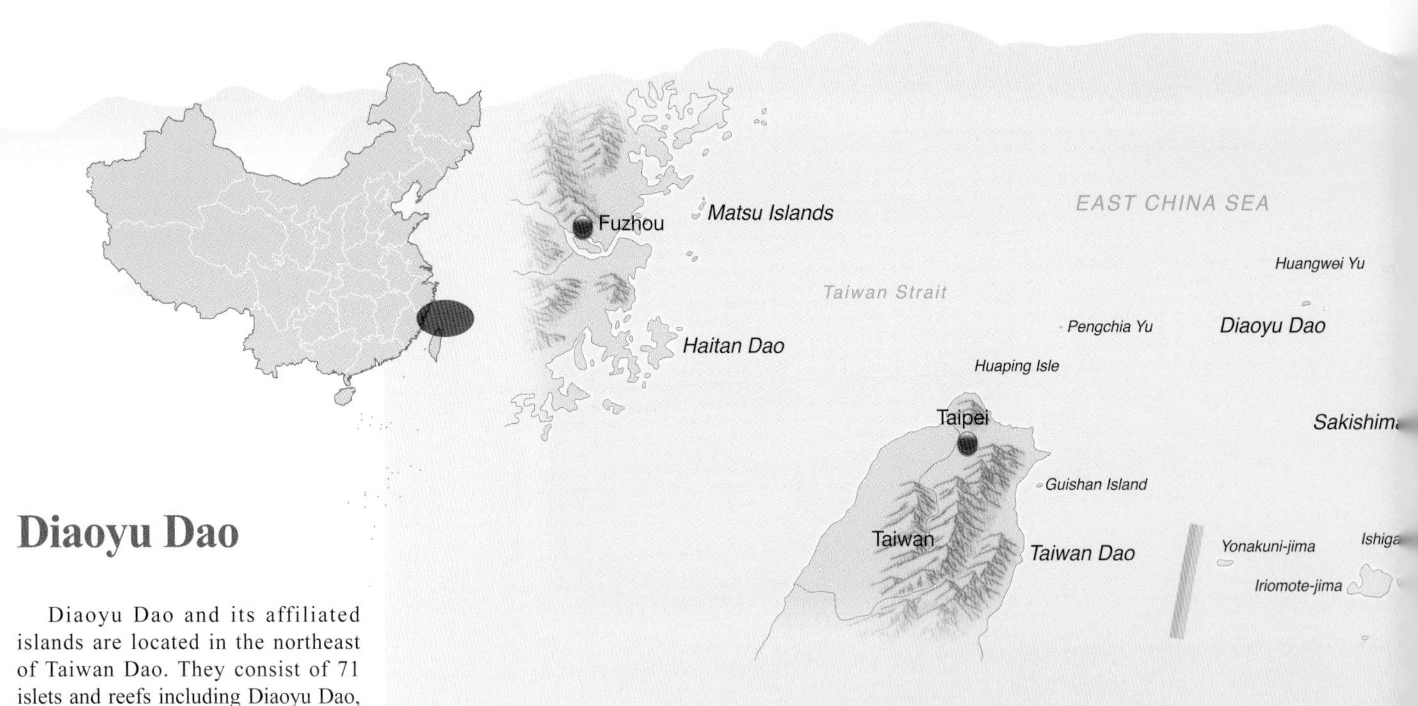

Diaoyu Dao

Diaoyu Dao and its affiliated islands are located in the northeast of Taiwan Dao. They consist of 71 islets and reefs including Diaoyu Dao, Huangwei Yu, Chiwei Yu, Nanxiao Dao, Beixiao Dao, Nan Yu, Bei Yu, and Fei Yu, covering an area of about 5.69 km^2.

Diaoyu Dao are in the farthest western region of the waters. Being the largest island in the region, it covers an area of about 3.91 km^2 and 170,000 km^2 of surrounding waters. It has a flatter northern side and a steeper southern side, with a mountain range running through from east to west. The highest peak, Gaohua Peak, is 362 m above sea level and located in the central part. There are also other peaks and four major streams.

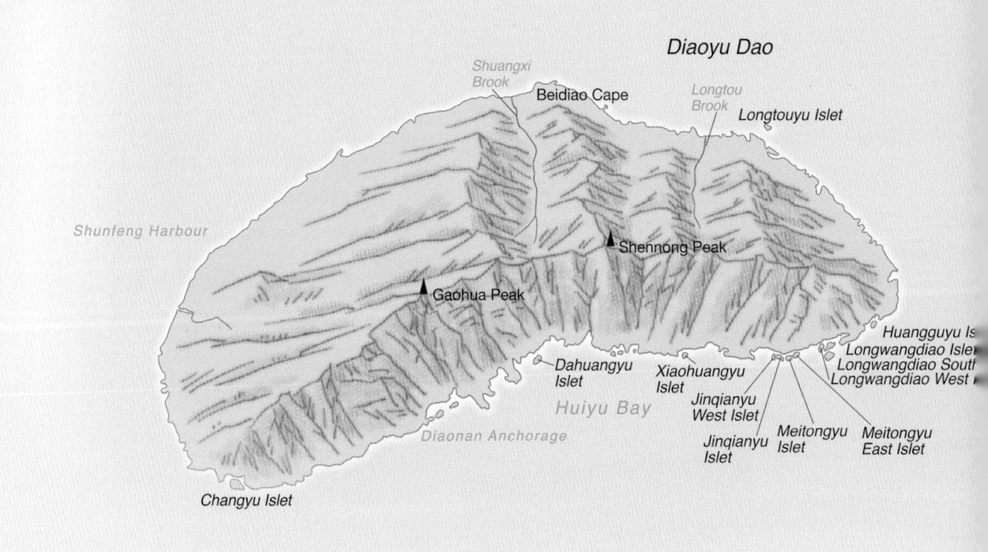

Strategic Value

According to *United Nations Convention on the Law of the Sea*, China and Japan are on opposite, non-adjacent continental shelves separated by the Okinawa Trough. Situated in the upper region to the west of the Okinawa Trough, Diaoyu Dao belongs to the continental shelf of China. The island and its territorial waters contain abundant oil-related resources and considerable economic value to various industries such as fishing industries.

History of Diaoyu Dao

The earliest record of the islands were found in *Shun Feng Xiang Song* written in 1403 A.D. (Ming). It was mentioned in several other records across Ming and Qing dynasties. For example, in *Recompiled General Gazetteer of Fujian* written in 1871, Diaoyu Dao was recorded as an important passing for coastal defence under the governance of Taiwan.

In 1894 after the First Sino-Japanese War, Japan occupied Taiwan and its affiliated islets under the Treaty of Maguan. In 1900, Japan renamed the Diaoyu Dao as Senkaku Islands.

In December of 1941, the Chinese government abolished all the treaties between China and Japan. On 25 October 1945, the Chinese government officially resumed its sovereignty over Taiwan and its affiliated islets.

In June of 1971, Japan and the US signed the 'Okinawa Reversion Treaty,' meaning that the US arbitrarily transferred the powers of administration of the Diaoyu Dao to Japan.

In March of 2012, the Chinese government announced the standard names of the Diaoyu Dao. On 10 September 2012, the Chinese government announced the baseline of the territorial seas of Diaoyu Dao. On 13 September, the Chinese government filed to the Secretary-General of the United Nations the geographical coordinates and marine charts, which contain the base points and baseline of the territorial waters of the Diaoyu Dao.

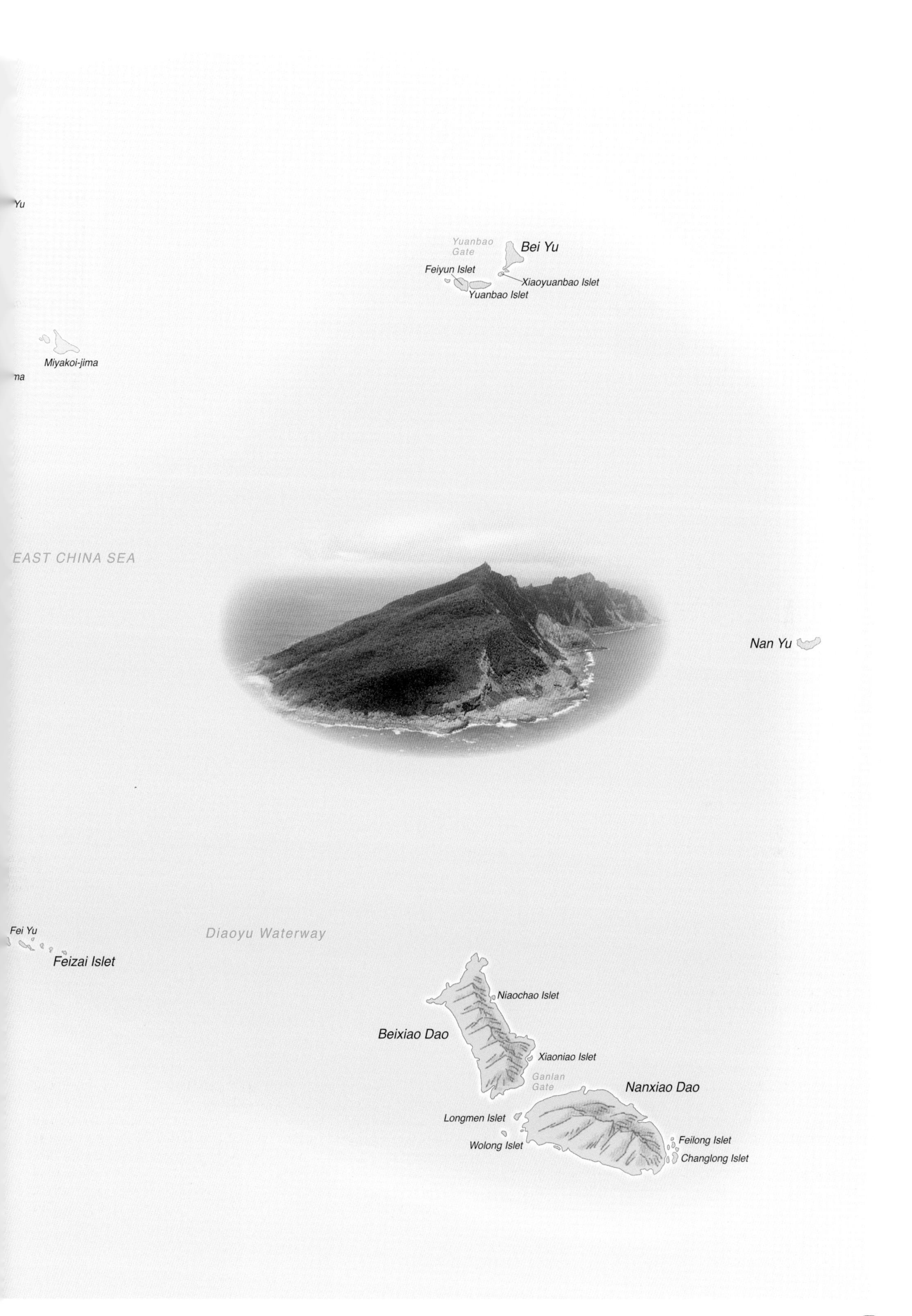
Yu
Yuanbao Gate
Bei Yu
Feiyun Islet
Xiaoyuanbao Islet
Yuanbao Islet
Miyakoi-jima
na
EAST CHINA SEA
Nan Yu
Fei Yu
Diaoyu Waterway
Feizai Islet
Niaochao Islet
Beixiao Dao
Xiaoniao Islet
Ganlan Gate
Nanxiao Dao
Longmen Islet
Wolong Islet
Feilong Islet
Changlong Islet

South China Sea Islands

South China Sea Islands are a collection of islands, sand bars, reefs, shoals, and sandbanks in South China Sea stretching about 1,800 km north-south from Beiwei Tan to Zengmu Ansha, and about 900 km east-west from Huangyan Dao to Wan'an Tan. It can be divided into Dongsha Qundao in the east, Xisha Qundao in the west, Zhongsha Qundao in the middle and Nansha Qundao in the south.

Nansha Qundao were discovered in the Qin Dynasty. Residence was built on Ganquan Dao from Tang to Song dynasties. After the Song Dynasty, they were called 'Qianli Changsha, Wanli Shitang.' Fishermen on Hainan Dao built over 200 routes in the region; a record book commonly used by the fishermen, *Genglu Bu*, recorded the names of 100 islands in the region. Zheng He from the Ming Dynasty also passed these islands during his voyage to Western Seas.

① Zhongsha Qundao

Zhongsha Qundao are 100 km away from the southeast of Xisha Qundao, in the southwest of Dongsha Qundao, and in the north of Nansha Qundao. Strictly speaking, Zhongsha Qundao are merely submerged reefs except Huangyan Dao.

② Xisha Qundao

Xisha Qundao comprises Yongle Dao and Xuande Dao. On Xisha Qundao, there are 14 ancient temples between Ming and Qing dynasties, plentiful stone memorials between the Qing Dynasty and Republic of China, and a historical site built between Tang and Song dynasties.

③ Nansha Qundao

Located in the southernmost of the region, Nansha Qundao are the most widely scattered among the South China Sea Islands. They consist of over 230 islands, reefs, sandbanks, and sand bars of a total land area of less than 3 km². The main islands include Taiping Dao, Zhongye Dao, Nanwei Dao, Zhenghe Qunqiao, Wan'an Tan, and Zengmu Ansha.

④ Dongsha Qundao

Nicknamed the 'Crescent Island' in ancient times, the Dongsha Qundao are also called Pratas Islands. They are at the hub of international shipping routes and under the administration of Gaoxiong City, Taiwan.

Yongxing Dao

Spanning 2.3 km², Yongxing Dao is full of luxuriant, tropical plants. There are facilities including offices, post office, banks, shops, a meteorological station, warehouse, a power station, a hospital, an island ring road, an airport with a 2,400 m runway able to accommodate a 737 Boeing jet, and a port that can accommodate a 5,000 ton ship. There are also flights and ferries to Hainan Dao.

Sansha City

Sansha City was officially established on 24 July 2012 ruling over Xisha Qundao, Nansha Qundao, Zhongsha Qundao and their waters. Sansha City covers 13 km² of islands and over 2.6 million km² of waters. It has the smallest land area, the largest total area, and the smallest population among all Chinese cities. The Sansha government is on Yongxing Dao, which is the largest among Xisha Islands and South China Sea Islands.

Monument to the Navy's Recovery of Xisha Qundao

Chinese Navy Landing on Yongxing Dao

The 'Monument to the Navy's Recovery of Xisha Qundao' is 1.49 m tall and 0.92 m wide, with its full name carved on the front. The signature says 'erected by Zhang Junran on 24 November in Minguo 35th year;' a phrase meaning 'Territory of South China Sea' is written on its back. After the victory of the Second Sino-Japanese War, many countries were preying on the South China Sea Islands. In September of 1946, the Republic of China sent Zhang Junran, the Captain Staff of the Republic of China Navy Command Headquarters, to lead troops into South China Sea. On 23 November, Zhang rode on the submarine chaser named Yongxing and arrived on the island, which was then named after the submarine as Yongxing Dao. A monument was built there in memory of the landing.

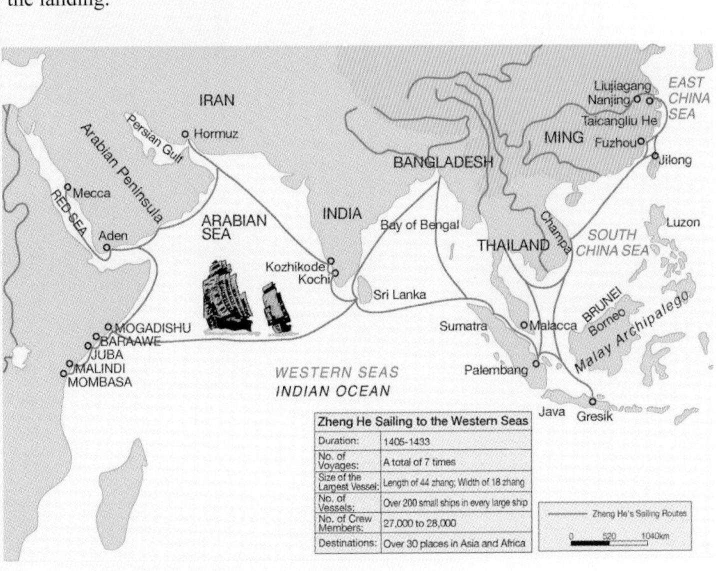

Zheng He Sailing to the Western Seas	
Duration:	1405-1433
No. of Voyages:	A total of 7 times
Size of the Largest Vessel:	Length of 44 zhang; Width of 18 zhang
No. of Vessels:	Over 200 small ships in every large ship
No. of Crew Members:	27,000 to 28,000
Destinations:	Over 30 places in Asia and Africa

Routes of Zheng He's Sailing to Western Seas

Zheng He Sailing to Western Seas

In 1405 A.D. (Ming), Emperor Chengzu appointed an eunuch named Zheng He to lead a strong fleet of more than 240 ships and 27,400 people to faraway places. They went to over 30 countries and regions in present-day western Pacific Ocean and Indian Ocean. The voyages were called Zheng He Sailing to Western Seas. He made a total of 7 voyages and stopped in 1433. As the fleet had to pass the South China Sea Islands, many of the islands and reefs were named after Zheng's voyages.